JUST DOWNLOADING

BY JOHN BALLENGER

ISBN 978-0-7414-7430-8 Paperback
ISBN 978-0-7414-7431-5 Hardcover
ISBN 978-0-7414-7432-2 eBook

Printed in the United States of America

Published May 2012

INFINITY PUBLISHING
1094 New DeHaven Street, Suite 100
West Conshohocken, PA 19428-2713
Toll-free (877) BUY BOOK
Local Phone (610) 941-9999
Fax (610) 941-9959
Info@buybooksontheweb.com
www.buybooksontheweb.com

Acknowledgments

For invaluable assistance in helping me prepare this manuscript, I give my heartfelt thanks and love to a number of people for urging me to write this book beginning with my wife Anne, my daughter Jacqui , and my sons Chris and Glenn. For years they have heard me from time to time talk about different aspects of my life and finally convinced me to put it to writing. A special thanks to my son Chris, who has spent an immense amount of his personal time helping me develop this book and aiding me in remembering portions of our family life that would have escaped mention without his friendly reminders and our mutual reminisces and laughs of those past years.

Thanks also to Kim and Ed Spear for their continued encouragement while I wrote and with special thanks to Kim for her creating the title to this book. The title, Just Downloading, which she thought of, speaks volumes to my long career in the computer industry. And a further special thanks to Ed, who worked with me and for me, for the extra special time he spent reviewing my drafts and adding valuable content to the years we spent together at C3. Also, an extra thanks to Tracy Vazzana, for her assistance in reading, correcting and adding any content to this book. In many ways this manuscript is also a story in large part of her husband John's career that spent seventeen years with me in building C3. Tracey's enthusiasm has been a large contribution in motivating me to complete my story. Her support and help has me feel like a young Ernest Hemingway in the making and I shall always feel grateful.

To all others including Ray Lunceford, Jim Kreiter (my brother in law), Jack Jennings, my sister Lorraine and brother Phillip for their support and assistance in reliving a number of years and tales that this book covers.

Contents

Prologue

Back in the Forties Jo Stafford, a popular singer of the times recorded a song entitled "Long Ago and Far Away", a melody that became an instantaneous popular hit. The lyrics, and music, even today, come back to me as fresh and lovely as they seemed that first day that I heard them, when I turned seventeen. "I dreamed a dream that day, and now that dream is here beside me."

Thinking back, I always dreamt and wondered what other people did, and who other people were That lived outside of my inner circle of family and neighborhood childhood friends I grew up with during the war years. Did everyone live like us? Not so, according to the books and newspapers I read, and the radio news that I listened to. I dreamt then, and knew that one day I would leave and venture out to meet these people, their cities, to see and taste what the magazines and novels touted as the "good life". I wanted to watch Joe Louis fight in Madison Square Garden, see the glittering glow of the Big Apple, with its beautiful and famous people; tinsel night life settings, with brightly lit up theatre Marquis! Watch the Yankees play in Yankee Stadium. To visit Hollywood, and meet the stars, Wayne Gable, Gene Autry, my old childhood hero cowboy star. What a blast that would be! I wanted to sail the Queen Mary to Europe, just like the Astors and Whitneys, and do what the other famous people of the time seemed to do on a regular basis. See a play in London, or climb the Eiffel Tower in Paris. Pretty amazing desires, at least to me, played through my mind back then. It was such fun to dream such crazy thoughts! Someday, just someday, I would do all that, be successful myself, and never look back!

Well, one day, I did leave that neighborhood and turned a number of my fantasy dreams into a reality, that deep down I never truly imagined would happen. So, this story you are about to read, to me, even today seems like a dream! My life, as it unfurled, reminds me of an act in a play that I wandered into by mistake.

For a number of years, my kids and others, including my wife, Anne has been nudging me to put these thoughts to paper-something to do with leaving my grandchildren a notebook on Granddad's life. So, I thank them all for pushing me into a writing mode and I hope these words you'll be reading will be interesting. And to my grandchildren, never believe you can't catch your dreams as you climb the tree of life; they are only a branch away!

"First Chapter"

I arrived on this earth on October 6,1931, in the early part of this Nation' depression, which lasted for most of the remaining thirties, the first of three kids, born two years apart; Lorraine, my sister, and Phillip, my younger brother, my father, John, known as "Mike," and my mother Eleanor Lorraine (Raymond). Both of my parents were born in Washington, D.C., third generation Washingtonians. We were born at the Columbia Hospital for Women originally the old Sibley Hospital, in North West, D.C., making us the next and fourth generation of Washingtonians. Not really a big deal, except when you grow up with virtually nothing, fourth generation status seemed to be worth something to be proud of. You see, when we came along in the height of the depression, my father was a baker with the Continental Baking Company in D.C., lucky to see a weekly paycheck of $25.00 a week; and there he remained for the next forty years, "eking" out a living, but providing for his family as best he could. He was a devoted father, who cared deeply for his family and did his best to see that we were taken care of. Not that that was unusual for in those years there were many like him; men and women that had little education, and had to go to work at an early age (my father at 16), to help support his own family, and shortly later a wife and three! As I look back at those early years and think of all of the successful, sophisticated and educated people I have met, worked with, socialized with, none of them come close to my father with a grade school education in terms of family connection and love, and work ethic. I can't remember his ever taking a day off from work because of being sick. Vacations, you've got to be kidding! He never took one, except for the weekend he got married. Couldn't afford to!

My mother I hardly remember! I'm told that she died when I had just turned five. Best I've been told is that she died shortly after my brother was born; some say because of childbirth, others claim it was contacting some sort of bird disease from canaries that she loved and raised. I really don't know; just one day she was gone and I was sent to live with my Grandparents Raymond; my sister went to

Figure 1: Jack with Mother Age 2-3

live with our Aunt Evelyn, a spinster, and my kid brother ended up with my father's brother, Uncle Bill and his wife Aunt Harriet. Kind of scary for three young kids, but we all survived and still surviving after some seventy years later! A little more on my mother, after all she deserves "equal billing" with my father. She grew up in D.C., the second youngest in a family of six children; George, Buddy, Edward, Alrose, and Evelyn. Her mom was Rose and father Alrose (Ray) Raymond. She grew up in Northwest D.C., at the time and still today, the "classiest" part of the city where the wealthy and top level government people lived.

Growing up in Northeast D.C., a less tony part of the City, I always remember my father's advice "when you grow up son, marry a woman from “northwest”! Anyway, my mom lived in Northwest D.C. all her life, went to public schools, and married my father before she turned twenty. He, you see, took his own advice to heart and married well! My aunts have been especially vivid and helpful in their remembrances of my mother as a very artistic woman who enjoyed painting, crocheting and was absolutely fascinated by birds. No doubt these interests were handed down to her by my grandmother Rose, who shared the same. A little on my mother's parents to share with you; my Grandfather was a successful businessman working for the Toledo Scale Company as their Eastern Regional Manager. You older readers may remember Toledo, they were the company that made the weighing equipment used in all the stores at the time to weigh everything from meats and vegetables, as well as by doctors, hospitals, or even local "guess your weight “and "win a prize" carnivals. They were everywhere-the computers of their day!

Although barely five, when I moved in with my Grandparents, I do remember quite a bit of the year I lived with them in 1936. We lived in a large home in Northwest D.C. for most of that year before my grandfather moved his family to Silver Spring, Maryland. My grandfather was an absolute tyrant, who ruled the household with an iron fist. My grandmother, God rest her soul, was just the opposite, kind, loving, the family peacemaker, who was the "go to" person when you had an issue with Grandpa. Little things come to mind: His favorite lounge chair by the fireplace, with its accompanying smoking pipe stand; the Evening Star newspaper awaiting his arrival from work. God forbid that anyone sat in that chair or dared to read that paper before he did, that could not and would not be tolerated and if you did you suffered a tongue lashing that "rattled your bones". The man had a temper, but it evidently worked in his business life, since he seemed to excel in that and worked for Toledo for many years. I was terrified of him as a kid, as were my aunts and uncles, but underneath his stern demeanor, there was a sense of kindness and gentleness that he showed to me versus others. He would often lift me up and place me on his lap, while sitting in his favorite chair without saying a word. I'm told that he was especially fond of and close to my mother and I suspect that the special care he showed for me was his way of expressing the love he had for her. I also remember my uncles George, Edward, Buddy and Alrose who lived with us. George, Buddy, and Edward were D.C. motorcycle policemen; "tough guys", and very imposing to a skinny little five year old kid; they were my heroes! I remember the joy of taking rides with them after work, sitting in the side car dreaming we were chasing down some gangster like John Dillinger. Heroes in my eyes, but in the presence of their father, courage and macho behavior vanished and they willingly complied with whatever the ole man wanted.

Country move

Sometime later, during 1936, my grandfather moved us to a new home he had been building in Silver Spring, Maryland. By that time my uncles had decided to stay in the city so my grandparents moved with me to the country. I loved the new house and the country, and especially my new dog, Seamus, a beautiful Collie. We spent our time exploring the streams, the woodland, catching frogs and trapping rabbits and squirrels. As a real five year old, almost six, I was real Daniel Boone, albeit without a coonskin hat. Evenings were spent listening to the radio with my grandparents, especially anticipating the next chapter of "The Shadow Knows" or "The Lone Ranger" or listening to President Roosevelt delivering a fireside chat from the White House, cheering us up and promising to lift us up and out of the Depression. Shades of today! How so much of today seems like that era then, except I feel today's problems are greater simply because of the larger and continued expansion of the population in the U.S. and the World in general, versus the smaller number of people living in the twenties and thirties. Societies are depleting and exploiting the natural resources of our planet, fresh water, forests being replaced by concrete and steel, staggering need for fuel, and housing, demand for food, expansive poverty festering and creating more civil unrest is the current outline for the future of the now more than 7 billion people that live on this earth, with MORE coming. In the end, the world's growing population will have to learn how to live better within its means. We're not going to find more fish; we're not going to plow more rainforest to create more calories. At Columbia University's Earth Institute, Prof. Jeffrey Sachs tells CNN "the consequences for humanity could be grim". In West Virginia, the Charleston Gazette editorialized about a "human swarm" that is "overbreeding" in a way that "prosperous well-educated families" from the well developed world do not. Back then, it was just a more simple life that I still yearn for! Anyway, I am beginning to stray into my own Malthusian ideas of what ails this planet today, and I must return to my more simple narrative, the story of my life, successes and yes, failures, many of them. Soon, however, my father had decided to remarry and bring our family, Lorraine (Rainey), Phillip and me, back together. So, again I packed up and started off into a new life, sad to leave my grandparents and Seamus, but excited to be back with my Dad, sister and one year old baby brother.

Chapter 2 – "A New Beginning"

And so a new start! My father married again to a beautiful young woman, from Maryland, Lorraine Eleanor Houchen. It would seem that back in those days the names Eleanor and Lorraine were popular as both my mother and step mother, as well as my sister, bore those names. Sort of reminds me of. George Foreman's kids all named "George"! Where and how my father met her is not known, although rumors exist that he knew her before he married my mother. Anyway, away they went on a quick weekend honeymoon to Atlantic City taking me along, I guess, for chaperone purposes. I'm not kidding; I have a picture of me taken on the Atlantic City boardwalk with them both, and me pulling a toy train! She probably knew by then what she was in for with two more kids; including an infant waiting for her once we got back. Well, we did get back and settled in a small cottage house that my father had rented in Forrestville, Maryland, a little town near D.C. I suppose he picked the town, since his brother Bill lived there along with my Aunt Harriet and their two children, George and Norma Jean. Also, my new step grandparents lived nearby. You might remember that I had earlier mentioned that my infant brother had lived briefly with my Aunt and Uncle after my mother's death. My step grandparents, by the way, turned out to be lovely people and welcomed the new kids into their lives. I used to spend an occasional weekend with them and grew especially close to my step grandmother, who was childless herself, and took special care of my brother, sister and me. Interesting, as did my stepmother, she raised four of my grandfather's children from a previous marriage.

Well, life started up again for everybody and what an adjustment for all. I remember especially what changed for me. No more Seamus to keep me company, and all of a sudden I was not the only spoiled kid in the room, there were now three and the new stepmother in our lives quickly made us "toe the line". She was an extremely serious woman; not especially warm to us, except for my brother Phillip, who at the time was under two years old. She was efficient however, and undertook running her new family as best she could. Meals were cooked, clothes washed, beds made, prayers at night made mandatory before bedtime, and so we took on as best we could, all the trappings of a happy family. My father continued to work at the bakery in the City and in his spare time at home built us a pigeon and chicken coup - and so, we raised pigeons and chickens for food and later on, a pet white rabbit was thrown in the mix for the heck of it. Thinking back, if my father had started building a boat, I would have believed he was adding animals and birds for the second coming of Noah's flood!

School started that first fall and I was enrolled in the first grade at Forrestville Elementary along with my cousin Norm Jean, who was my age. It was then I was assigned my first "grown up" responsibility, watching over Norma Jean on our daily walk to school. It turned out that she was subject to seizures from time to time and I had to carry a wooden spoon to place in her mouth when she suffered one. I learned early in life how to be a paramedic and performed that duty many times during that year. I loved school; turned out to be a quick learner, and looked forward to it each and every day. The only downside at the time was my own health. I was frail and skinny, and developed a very bad case of asthma since my mother had died, and from time to time I would have major attacks where I could hardly breath, that would cause hasty trips to our local doctor for a shot of adrenalin, the only therapy at the time to treat asthma. I can remember feeling very guilty whenever a trip to the doctor was necessary seeing the look on my father's face. He was desperate for me to get treatment, but every trip and treatment cost $5.00 and when you only made $25 a week, it played havoc with trying to feed a family and pay the monthly rent. Luckily, several years later I was enrolled in an

Figure 2: Jack, Phillip, Rainey

experimental asthma program at the old Georgetown University Hospital, when a new drug spray for asthma was being used. It was successful and I carried and used the spray into my late twenties, when remarkably I outgrew the allergy. On a humorous note, at the time, not funny but embarrassing, I was on my honeymoon with my wife, Pat, and on our first night in Florida; she had to rush me to a local hospital for a shot of adrenalin for an asthma attack. Never thought on your honeymoon night one would need a shot of adrenalin! But it worked; we were together for twenty years.

Chapter 3 – More Moves

Figure 3: Brother (Phillip), Father, Stepmother, Sister (Rainey) and Jack – Circa 1939

But back to my childhood - After one year my father decided to move from Forrestville back into the city to be closer to work and so we moved into a small second story apartment in Northeast D.C. At 5th and G Street, a building that still exists today as a rundown night club. This area of Washington was a hot bed of commercial activity with one of the two central markets serving Washington residents, directly across from where we lived on G Street. On weekends especially, the area was alive with people shopping, the sounds of music, and a social meeting place as well. We lived in what then was an Italian community just two blocks away from Chinatown, so that within a five minute walk, one could envision he had just visited a couple of foreign countries. Chinatown, and what was known as Little Italy in D.C., still exists today in those same areas of the city. My step mother's Aunt Helen lived just two blocks from us and we would often visit her. She was very old when I first met her, and the thing I still remember about her, was she was bald and wore a terrible loose fitting black wig. I must remember to tell you a story about Helen and her wig that my sister and brother still find hilarious, whenever it comes up in conversation. I'll get to it later!

At any rate, we settled in the new apartment and it was time to think about enrolling me in a new school. I was ready for the second grade and being Catholic, my parents selected a little Catholic school, St. Mary's, just two city blocks from where we lived, in the heart of Little Italy, at 3rd and G Street N.E.; taught by Nuns that had learned torture techniques directly from the CIA! And there I went! As it turned out, I was there for a year before my father got restless and wanted something better for the family; so we moved again, further north in Northeast D.C. to the Brookland area of the city, thanks to a "house loan" from my step grandfather Houchen. Brookland, in those days, was never Washington's most fashionable address, but it attracted a steady stream of middle-class families eager for it's shady streets and single-family homes. Smithsonian scientists hunted for archeological artifacts from earlier colonial times here, after they got home from work. Professors from Catholic University lived there. Brookland was the home for many Catholic institutions, including the Franciscan Monastery, various catholic nun convents as well as Catholic University, Trinity College and the Immaculate Conception Shrine. Catholic University was also the home of the famous Father Harke Theater where many students went to study theatre and from there went on to hopefully star under the lights of New York or Los Angeles stages. Names such as Walter Kerr and his wife Jean graced the CU stage before they moved on to New York, where Walter became an acclaimed theatrical writer and critic. Writing and directing musicals like "Goldilocks" with his wife Jean. Also, the widely acclaimed Frances Sternhagen, thrilled audiences at the Old Hippodrome Theatre Stage at 9th and New York Avenue. It later became the Arena Stage and before she won national acclaim as "Daisy" in the New York stage hit "Driving Miss Daisy". I lived there doing that era; actually met many of those people when I got older, and it was where I developed my love for theater to this day. Quite a number of Howard University faculty lived there as well, including Ralph Bunche who chaired Howard's political science department and later became the first African American to win a Nobel Peace Prize. Dwight Eisenhower reportedly once asked Bunche how he liked living in Washington. Bunche said, "he liked it just fine, except he had to send his kids across town for school, there being no black schools in Brookland" at the time. Lucy Diggs Stowe, the first dean of women at Howard and a founder of the Alpha Kappa sorority was a Brookland resident. Brookland was a home for many academics and theatrical and journalistic talents. John Preston Davis who published Our World, a pre cursor to Ebony and Jet, was a Brookland resident. The Crosby Noyes elementary school two blocks from where I lived was named after the publisher of the Evening Star newspaper, at one time one of the major newspapers in D.C. Marjorie Kinnan Rawlings, a contributor to the Washington Post children's section and author of "The Yearling" grew up on Newton Street in Brookland, a short walk from the Newton Theatre where as kids we watched movies every Saturday afternoon. I remember those weekend movies at the Newton; my mother would give me twenty-five cents for the Saturday afternoon show, which paid for the movie (ten cents) and fifteen cents for candy. Always great entertainment! Show included a "movietone newsreel", movie, cartoon, and your favorite serial short that was shown from week to week, somewhat like a daytime TV serial show today. On a lighter note, I had to earn the twenty five cents! What the job entailed was also taking my sister to the Newton as well. I would be furious with having to babysit my younger sister when I wanted to spend the time with my friends. She laughs now about those times, but she was not a happy gal when I would make her sit two or three rows behind me in the theatre.

Ours was a typical Washington row house located at 613 Franklin Street, seven or eight blocks from Catholic University and Trinity College, then an all woman's college. We settled in and I was enrolled in St, Anthony' Elementary school, beginning with the third grade. St. Anthony's was located in the very heart of Brookland at 12th and Monroe street N.E. and like most catholic schools at the time, was connected to the local parish church, St. Anthony. Also, just as a side note, there were many private catholic schools sprinkled around the D.C. Area and it was Church policy that the majority of elementary schools were coed thru the eighth grade, but once you reached the "puberty “stage of life all of the catholic high schools were separated by sex; girls went one way, boys the other. Except for one! ST. ANTHONY'S. God must have personally directed my parents to Brookland, because every catholic kid in the city wanted to go to a coed high school, and I turned out to be one of the "lucky" ones.

So! Enrolled in a new school, my third in three years, I set about adjusting to my new surroundings, meeting new classmates, making new friends and coping with a new order of nuns, the

Benedictines, a catholic teaching order. In those years most all, if not all, of the children going to city catholic schools were Caucasian, a mix of Irish, Italian, German and French. Many were first or second generation children of blue collar workers; plumbers, carpenters, electricians, small business owners and government workers. The Catholic schools, at the time were ranked highly from an academic standpoint; focused on the 3 R's and obviously teaching basic catholic doctrine and values. School uniforms were the order of the day and rules were rigid; rulers across one's knuckles for misbehavior in classrooms were in vogue and school conferences with parents over poor academic or behavior problems were commonplace. You dreaded carrying that note home to your parents from Sister Rosario requesting an audience with your teacher. An hour per day was spent in class being taught Catholic catechism, with much of the balance being taught math and reading, with special emphasis being placed on penmanship and spelling. To this day I swear I can look at a person's handwriting and say, "you must have gone to a catholic grade school". School lunches were unheard of- you brought your own lunch box with the school providing milk each day, as long as your parents could pay the 25 cents per week for the drink. I still remember Friday lunch boxes. Canon law back then precluded one from eating meat on Fridays, so my lunch box sandwich was always tuna fish, or more often peanut butter and jelly. To this day I cannot stomach peanut butter and jelly sandwiches! If you believe in an afterlife, Heaven and Hell, and you unfortunately end up in Hell, and then peanut butter and jelly have got to be your major punishment for all eternity! Most kids walked to school from where they lived usually within a radius of five to ten blocks. You walked, whether it was cold or hot, rain or snow! In my case, (and later my sister and brother) the walk was eight to ten blocks, over the Franklin Street Bridge, past the Stone Straw Factory (where straws you use when you sip a drink were made-it's still there), left by the Crosby Noyes public school, across the Monroe Street Bridge, and then several seminaries and convents later school appeared. All in all about a thirty minute walk. Along the walk, I would meet fellow classmates on the way as well. Kids like John Falcone, later to be Father Falcone and later yet, Monsignor Falcone in the Washington Diocese. Others like Jack Jennings, still a good friend, living in the Northern neck of Virginia, as a retired Navy jet pilot, Vivian Hudson, Betty Houlihan, and Dick Goetzinger, and many others like Dick, a good friend, who is still active practicing tax accounting. Mentioning John Falcone reminds me of a unique way that catholic school classrooms were organized in those days. There were usually up to 30 kids in a class sitting in old fashion" slide in desks" arranged by academic standing. So, if you were the smartest one in the class, you sat in the first seat to the left of the classroom and the dumbest, the last seat in the last row to the right of the classroom. Falcone was always in the first seat and I in the second seat all the way thru 8th grade when he left for the priesthood. I used to always pray and hope that God would give him a calling before the 4th grade so I could move up, but he never did. I remember my grandmother Rose wanted me to be a priest. She prayed and prayed for such a vocation, but evidently, God heard the Falcones! Monsignor John was one smart guy and a close friend all through grade school. I can only imagine the out roar from parents today, if classrooms were setup that way- "you're discriminating against my dumb kid"! I'm suing the school system!

Chapter 4 – Early Times

So by now dear readers you are probably bored stiff reading about catholic school diatribe and asking yourself, "is he ever going to move on from nuns and priests."? Well, yes I am, but you have got to know I'm going back there because there is much more to bore you with before I finish.

Meanwhile, before you knew it my sister and brother were attending St. Anthony's grade school, and I was graduating from elementary on my way to High School. There is something special about going on to High School for a kid; like, turning sixteen; first date; first kiss; getting a driver's license; first car; maybe your first beer, at any rate it's very special and exciting. All of a sudden a grownup in your mind- That was me! Besides, I had something special happen to me. I had taken a citywide exam for a full academic scholarship to St. Anthony's and was one of three persons, including John Olsson who became a lifelong friend, and Vivian Hudson, a fellow classmate to receive one. Olsson had a brilliant mind; came from a large Irish/Swedish family of which a number were Mensa gifted. John lived in Northwest Washington and moved to N.E. to attend high school and after a stint in the Army as a Russian language instructor. He graduated from Catholic University with a degree in Music and went on to work briefly for a store chain, called Discount Records, on Connecticut Avenue and shortly thereafter started a chain of book stores in the Washington area called "Olsson's Books & Records which he owned and ran until shortly before his death in 2009. His stores competed mightily for a good number of years against the giants of that industry, but sadly closed their doors in 2009. Unfortunately, that has become the story for too many boutique or neighborhood bookstores since the advent of megastore or e-book web sites. John and I remained best friends until the day he died, and had much in common, except for politics! Theater was a common denominator between us, and we, along with other friends, were always finding ways to see the latest plays at local colleges, neighborhood theaters, and when we could afford it, taking a Greyhound bus trip to New York to catch up on the professional theater, buying the least expensive tickets we could, eating at the old Horn & Hardart cafeterias located throughout the city and bedding down at the New York YMCA for fifty cents a night for a room, and walking down your floor hallway for bathroom and shower facilities. Not always a "safe" trip! We met an awful lot of interesting characters at the "Y", many who lived there full time, including many out of town thespians, or would be actors, looking to make it "big time" on the New York stage. In D. C. we found part time jobs at the National theatre as ushers; pay was lousy, but we were able to watch the shows and occasionally to meet the performers at after show get togethers.

The Horn & Hardart automats were popular meeting places for all. They were an early combination of fast food eateries with coin operated machines dispensing food; women with rubber tips on their fingers, "nickel throwers", as they were called, giving customers five cent pieces required to operate the food machines, in exchange for larger coins and paper money. They became great equalizers because paupers, investment bankers, out of work actors, or a famous actress might sit together at the same table eating and sipping their coffee. Coffee at 5 cents a cup was known as the best in town and over 90 million cups were served each year. Irving Berlin, the composer, of "God Bless America" wrote a famous song about this delicious brew, "Let's Have Another Cup of Coffee" which became the store's theme song. Sadly, the last of these stores closed in the seventies and were replaced with Burger King franchises.

Also, during those years, I was fortunate to get my first part time "paying" job on weekends working for the old, now long gone, Kann's Department store at 7th and E Street in downtown D.C. There I worked for a middle age single woman by the name of Miss Sigman, who was the vice president of business administration, running the business operations of the store for the owner Saul Kann and his family. What a great job! And it paid seventy five cents an hour. Miss Sigman had never married, had no children and doted on me as if I were her only child. Whenever she could, she'd stop and chat, and inquire about school, what I planned to do after high school; was glad I hoped to go to College, gave me an occasional raise, and sort of became a second stepmother, I remember that she asked me once, after giving me a ten cents an hour raise, what I hoped to do when I grew up. I told her I always wanted to be an architect and hoped to make as much as $200 a week. I remember the smile and she saying, "Jack, I know you'll be successful designing buildings, and I betcha you make a lot more than that!"

Also, I met my first love at Kann's- a young gal by the name of Linda, who like me worked part time while going to high school. She was a "knockout" and at 16 we were both talking marriage after the first week! After all, we were both employed, in love, making a combined $1.70 per hour, and only one year away from high school graduation-we had everything we needed to make it work, except for one thing!

I didn't have a car, or a license (until my senior year) and she lived in Tuxedo, Maryland, a one hour bus drive from D.C. Needless to say, distance killed the "she's my one and only ever" love of my life. But we did keep in touch and in my freshman year of college, she invited me to her wedding in Tuxedo. It was one of the saddest days of my young life. I went and after the wedding party sat in the church for over an hour, the groom evidently got cold feet and never showed up. I felt terrible for her, slipped out of the church, and never heard from her again. She was gorgeous and bright and hopefully she found "MR, RIGHT" later on.

"Social Life in the City"

Slipping back in time, as I write and reminisce about this time period, I would like to talk a little bit about how we and other average families lived and spent our time together and with others. Other than work and school, families socialized mainly with close neighbors on your street block and various relatives from around the city. Entertainment and travel were limited, if not because of lack of money, then basically because there was not much available for the average working-class family to do. One of the most popular pastimes was Saturday night card playing, (pente-ante poker) and drinking, that would go on into the early hours of the next morning, around the family dining room table. Aunts and Uncles, would arrive with my cousins for early dinner; the table would be cleared for card playing and the children were sent elsewhere in the house to listen to favorite radio programs until bedtime called; where cousins and all would double up in our beds until their parents called the evening quits, and toted everyone home. I can't recall ever hearing of, or having a babysitter. That profession was not created until years later. Other than weekend card night visits, I don't recall our taking any vacations except for visiting other family members, and that was usually on weekends, mainly in the summertime. Occasionally in the summer, my parents would take us for day trips to one of the local beaches on the Potomac River. Pope's Creek, Md. was a favorite, located on the route John Wilkes Booth followed on his attempt to escape after assassinating Lincoln. Right on the Potomac River across from Virginia, it was an idyllic daytime trip to escape the heat of the City and enjoy a day of swimming and eating Chesapeake hard-shell crabs at Captain Robinson's crab house. My father's cousin, "Aunt Georgie" owned a summer cottage in Colonial Beach, Va., on the Bay close to the City and my brother and sister, and I, would sometimes be invited to spend a week with her. Now that was heaven! No parental supervision except for "Aunt Georgie" who was a "hoot", and let us do anything we wanted. Later on during my high school years, kids that could drive, would hop in your "buddy's "car and head for North Beach or Cedarhurst on the Bay in southern Maryland. Oftentimes on a summer weekend Saturday, kids would ride the streetcar from the City out to Glen Echo Park, Md. the City's local "Coney Island", for a day of swimming and riding the roller coaster and other daredevil rides. One of our high school friends, Glenn Saffran and his family owned a cottage in Cedarhurst and we spent many a summer weekend there with other high school friends from around the city, friends like Bill Craven, later a high level Government executive in HEW. Paul Barrett, a later successful owner of a large D.C. Construction company, who's uncle was head of the entire D.C. Police Department, Frank McGlister, the owner of a major insurance agency in the area, who later handled all the insurance needs of my various companies, John Olsson, talked about earlier, Mike Miatico, a later multi- millionaire business owner, who was the only kid who owned his own personal car in high school, and a convertible to boot. Guess who got all the women! Ocean City, Md. became a favorite beach spot, where you had to take a ferry to get across the bay, years before the Chesapeake Bay Bridge was built. Six or seven of us would pile into a friend's car on a Friday weekend; get to the

Ferry and before crossing the Bay and paying the individual fare, two or three of us would jump in the car trunk, to escape paying a fare until we got to the other side of the water. Weighing only 90 pounds at the time, guess who was always one of the first that was forced into the trunk!

Everyday life in the city evolved around one's neighborhood- your friends, your family, schools, play, and even shopping. Back then, there were few large grocery stores, and most shopping for food was done at your neighborhood District Grocery Stores (DGS). These small stores were located throughout the city and almost everyone shopped there. They were the mainstay for D.C. grocery shopping for over 50 years in the City. An interesting side light of this topic was years later, I moved to Potomac, MD in the 1980s' and the last DGS was still operating in the Potomac Shopping Center. You didn't use cash or credit cards, the owner simply kept a little sales book with your name and charges and you settled up with him at the end of the month. Later on, somewhere in the late 1940s', the Sanitary Grocery stores, which was the forerunner of Safeway came to town, opening more central city larger stores and as transportation improved in the city, refrigerators, in homes, became larger to hold more produce, and they gradually took over from the DGS smaller stores. The era of "ice boxes" and daily home delivery of milk and ice, and portable electric fans for home cooling, was coming to an end, and being replaced with window air conditioners, milk now sold in paper cartons in larger stores, ice being miraculously created within refrigerator freezers, and larger "Ronald Regan" GE products such as refrigerators and ovens delighting the housewife. Party line telephones were also slowly being replaced with direct dial rotary phones. No more waiting for your party line neighbor to hang up or plead with him "to get off the line" so that you could make your call. Lillian Tomlin type operators were also becoming "extinct". Later, in the early "fifties" came the first RCA television set to replace your old Philco or Emerson Radio. I remember our first tiny five or six inch black and white set. What a miracle! And, later that decade, color TV, the Jetsons, cartoons to keep the kids occupied, Sid Caesar and Imogene Coca for the parents. Dick Tracey, a comic strip detective, with his "cell phone" watch was just around the corner to be replaced one day with the creation of Android and I-phone technology. How much better could life get?

As it turned out, the Franklin Street neighborhood, where I lived, provided a lot of other friends that attended public schools in the area, many that are still close friends to this day. We played sports in the alley streets behind our homes, apple crates nailed to telephone poles as baskets for basketball games; trash can lids for baseball bases. Most of us belonged to the local Police Boys Clubs and played softball and football in leagues at the local playgrounds, places such as Turkey Thicket, in Brookland, and the Edgewood Playground, near my home. Friends like Ray Lunceford, who lived on Girard Street, a block from me; went to McKinley Tech High School, and later met up with me at the University of Maryland; graduated, served in the Army, and went on to co-found and run one of the largest public technology companies GTSI in the Washington Area. Frank Wright, who went to Eastern High School in N.E. has been a dear friend for years, lived close by; a scholar and artist; graduate of George Washington University, and American University; lived and through scholarships and grants, studied art and painting in Paris and Florence for several years. For years, he has been a recognized artist, noted for his historical paintings of the civil war; with his works in the hands of many well-known collectors throughout the Country, including the Kennedy Gallery in NYC, and many corporations, such as Capital One, the Charles E. Smith Companies and the B.F. Saul Corporation, commanding commissions for his works upwards in the six figures. In addition, he currently is a professor at GW, teaching art to University students where he has been teaching for over fifty years. I can remember sitting with him on my Franklin Street front porch, where he would show me some of his early work, just a kid then, with dreams, that came to fruition. Other friends like Charlie Offutt, a fellow classmate at St. Anthonys, who lived directly across the street from me. Charlie was a very quiet guy who was brilliant in school, an outstanding athlete, and an absolute terror to watch in a teenager fist fight. He wasn't called "knockout Charlie" for nothing. But Charlie's fame was not made on the streets, with his fists, he went on to graduate with a PHD from the University of Maryland and taught school at John Carroll High School in D.C., as well as working as an assistant football coach under the famous and legendary John Wooten, a local football coach. It was during his teaching career that he developed an educational program to assist NBA players in developing and rounding their lives beyond basketball, including how to manage their business and financial affairs. The program became a must for players in the league and officially promoted by the NBA. It is often said that the importance of one's success in life and contributions to life can often be weighed by the size of one's obituary in the papers, and if so, then Charlie's was exemplary. His was almost a full page in the Washington Post, when he died in 2010. Also another close friend, Jack Jennings, who lived on Tenth Street, N.E., along the walk route I traveled to school. Jack was an outstanding athlete in high school, went on to graduate from Catholic University with a degree in electrical engineering, joined the U.S. navy and then flew fighter jets for the U.S. Navy before he retired to the beautiful Northern Neck of Virginia. We still keep in touch and share times together talking over how we both have lived during some of the best 50 years in the world, enjoying and reveling at some of the most exciting developments in history: advancements in medicine, i.e., open heart surgery, elimination of polio, malaria, magical pharmaceutical drugs to cure, eliminate or minimize a plethora of diseases and infections. In technology, a world I made my career in: the era of computers, once the size of a Wal-Mart store, now downsized to the size of your hand, the cell phone and the Internet, fast replacing the need or use of telephones, bookstores, newspapers, and yes, shortly the need for your post office or postman delivering mail to your home - jets, rockets to the moon, satellites, all such things that have been created mostly since the early 20th century. And then of course, the little things, credit cards, ATM machines, bottled water, frozen TV dinners, remote controls, penny loafers, cheese in a can, and even "wine in a box", that have made our lives so much more enjoyable, or perhaps lazier. Or on second thought, have they really? I'm sure many of you have second thoughts about all of these wonderful inventions when you're sitting at dinner with your kids or grandkids when the only conversational sounds that you hear are the quite sounds of a cell phone txt being sent, or a game boy being played. One doesn't use his or her voice box any longer- voice boxes will probably be DNA eliminated in future generations, or perhaps like an appendix, you'll be born with it, but it really doesn't serve a useful function. All I know is, if I call one of my children or grandchildren, they will never call back, but if

I txt, there is instantaneous response. Whatever has happened to real conversation amongst us today?

Oh, and I almost forgot to tell you the story about my ole Aunt Helen. Holidays like Thanksgiving and Christmas were always big family days, just like today, and my parents always had a ton of relatives over for dinner celebration. Well, Aunt Helen being a widow and quite elderly was always there for Thanksgiving, dressed in her best 1880 finery, including a well combed black wig, which honest to God, always looked to me like she put it on backwards. You all know kids! So naturally, we all would watch her from different spots around the table and wait for something funny to happen. You could always expect something hilarious for kids to laugh at, and sure enough she did not disappoint! While eating her soup, she leaned over and the wig went into her soup bowl. Well, you can imagine how we handled that. But that wasn't the end- she picked it up out of the soup and put it back on her head! Needless to say, poor ole Aunt Helen never came back for Thanksgiving dinner again. And come to think about it, I don't ever remember getting a Christmas present from her that year. Wonder Why?

Digression is so wonderful when you are the story teller! You can write all you want, in any way you want, and mix up scenes, passages of time, story content, exaggerate the tales, confuse the reader completely and suffer no consequences whatsoever, except for one thing! The reader doesn't have to finish the book! And so, I hope you will bear with me, be patient, and hope that by the ending there is a coherent story for you to enjoy and equally important a story that you will understand and like, and finish! Life is such a saga for each of us. One would think that God could have found a better use for his time than creating it. Just think about it for a second, when we have a child, or more join our family group, there is always much joy surrounding that event, but then all hell breaks loose as the years rush by; the ups and downs of life can be both exhilarating but also exhausting. Why does He take us thru all of that? It has to be a lot of work for Him, but in all fairness to Him He does push us thru life in the fast lane, because life is short, and maybe that's the way He planned it. That way He doesn't have to deal with each of us for too long on earth. Where am I going with this, I don’t know; it sort of ties in with the ups and downs of our individual lives and what we do and make of it in a short period of time. I sort of think of life like a child who sees a Trampoline ride at a local park, finds a way to climb up on it, jumps high in the air, tumbles back down, jumps and bounces to another area, falls, hurts himself, cries, is helped to his feet by another, jumps again, laughs and lands someplace else and continues to succeed and fall, jumping on this thing until he gets tired and decides to get off. That's sort of how our lives are lived. Unlike, say a merry- go-round at that same park. Merry-go-rounds are sequential and timed. You know they only go one way, and that they only last for two minutes. Life is unpredictable- you don't know where it's going, what's going to happen or how and when it's going to end.

God is always trying to surprise us. Maybe that's how he maintains his sanity, or better yet his humor. So I guess that's why I'm into digression at this point; you'll just have to live with the back and forth jumping around of this story until the end.

And so, back to my finishing up my grade school years in 1945! Can you believe it- that's over 65 years ago! I should be writing this in a "retirement home" by now. Elementary catholic school was then eight years long before you went to high school for four years, whereas public schools were six years of elementary education, three years of Junior High and then three years of high school. Why am I telling you this? It's really probably not of much interest, but it was just that as school kids, we were "pissed" because as we reached the 7th and 8th grades still wearing knickers and long stockings, dressed as children still, our public school friends had moved on to junior high, wearing long pants and short dresses, enjoying early stage dating, proms and all that other grown up stuff. But time moves on and so I graduated from grade school and jumped from knickers to long pants. Most of the kids that were in my grade school continued on at St. Anthony's, but a few left to go to other catholic and public schools elsewhere in the city. If your parents could afford the tuition, you might end up at St. John's military or Gonzaga High, the two most prestigious catholic schools in the city, both academically and athletically; they both had powerhouse athletic football and basketball teams and recruited the best athletes from the area. Many famous and well known people went to these schools. Pat Buchanan, who most of you have heard of, went to Gonzaga, a Jesuit school. Pat became a very famous speechwriter for Richard Nixon and later ran for the White House himself, but lost the nomination to George W Bush. Since those days, he continues to write political columns for various publications and is a popular commentator on several national and local television shows. My brother-in law, married to my sister, Lorraine, went to Gonzaga as well, just a few years behind Pat. One of the most talented high school basketball and football players in the D.C. area was Jack George, the Larry Bird of his time, went to St. John's. George broke all high school records for the School in the late 1940's and was picked as an "all Prep" player by all three major newspapers in the City. Dave Addie, sports writer for the Washington Star (later married to Pauline Betz, the famous female tennis pro and Wimbledon champion), called George "the best high school athlete he had ever see perform". Morrie (Mo) Segal, the sports writer for the Washington Daily News "dittoes" the same! I later met Mo thru an introduction by a close friend of mine, Foster Shannon, President of Shannon & Luchs, the major real estate firm in Washington at the time. A little more on Foster later, he was a beloved friend and confident of mine for many years. Segal, who became a friend, was a character known locally by everyone. He was a natural born raconteur who spent his nights at Duke Zeiberts restaurant, a local watering hole, entertaining friends and talking sports and politics. I don't remember him ever going home at night! Thirdly, Shirley Povich of the Washington Herald, who wrote elegant and knowledgeable sports columns for years, and later at the Post, that were syndicated everywhere, also proclaimed George as "the best ever"! Povich was truly the "darling" sportswriter for America for well over a half of a century. The Herald later became part of the Washington Post, the morning paper and the Daily News merged with the Washington Star, the evening paper in D.C. The News had been an afternoon paper, but was published only five days a week, but was extremely popular with Washingtonians. Well known newspapermen wrote for the News, including Ernie Pyle, who was a corresponding editor for the paper, and a famous war correspondent. Also, folks like Judy Mann, noted for her Vietnam protest at Columbia University; Sam Gordon, who so much believed in UFO's swirled our imaginations with his crazy stories. Bill Beale, the chief photographer, for the paper, won many awards for his work and had many photos that appeared in Life Magazine and other publications. Oh, by the way, I need to remind myself to tell you a little story about how I lost my "virginity" while serving as a paper carrier for the Daily News when I was only 14, but I don’t know if I should or not. It's a story that only a few old neighborhood buddies had heard years ago. My sister will be shocked when she reads it! But back to newspapers, for a moment more, they were a very important part of one's daily life. You read the Herald in the morning over your coffee before work, and later the Post; the Star or the News in the evening. In addition, there was an advertising paper that was circulated every Thursday to all households in the City, called the "Shopping News". Carriers, like me, were paid 75 cents a week to deliver it. It was always chock full of items to buy from local stores and was heavy as hell to deliver. Still have scars on my shoulder from carrying the "load".

Chapter 5 – "Teen Years"

High School was new and exciting. Different buildings, different teachers for different classes, (unlike grade school), and different classmates since many of the newcomers came from all four areas of the city. But what didn't change was the discipline in the classroom, school uniforms for the girls (blue for school colors) and white shirt and blue ties for boys. Religion classes were often taught by the local priests, and you could sense and feel the added emphasis placed on religious vocations and the need for serious reflection on the part of students to consider this as a career option. Some did, two women I especially remember, Alice Ricci, an Italian hell raiser, who would be the last on earth you would think to join a nunnery. But she did and Roberta Wilson, now retired in New Jersey, who recently wrote a little remembrance piece for the school newspaper, recalling all the great and fun times she had there. The school was small, consisting of around 150 in all four class years. Our class of around 35 divided equally between boys and girls. As in any school, you meet certain people that you socialized with, and. together with a "selected rat pack" of guys we rolled thru four years of high school learning to drink beer, smoke, and puzzling over women while worrying as to what we would be doing after high school. Much of our social life in those days centered on the weekend sports and catholic social dances sponsored by different parishes around the city. In that generation, you were always being told by the nuns to pray for a vocation and barring one to be forthcoming, you were pushed towards at least dating a Catholic and "marrying your own kind"; henceforth the popularity of these Parrish dances. The biggest and most popular of these weekend dances was held at St. Martins parish church in Northeast Washington. Kids from all over the city would flock to St. Martins Hall on weekends. There was where you heard the best bands and met the most interesting dates. Since few of us had cars, we all traveled by street car and bus and it was always a challenge figuring out how to escort your newly found "love" home from the dance and then to find transportation from her house to yours. Many a Saturday night turned into an early Sunday AM walk because I missed the last bus or street car.

It was not until my senior year, though, before I met my first true catholic girlfriend. Introduced on a blind date, I immediately fell in love with Marilyn Powers, a student at Holy Name Academy, a girl's school in Washington. We dated for most of that year, but by then I was interested in college and she was interested in marriage, so we broke up as I was heading toward freshman year in college. Senior year for me, like for most was fun and exciting. Since I was sixteen at the time, a little younger than most in my class, I didn't learn to drive until my senior year and convincing my parents to lend me the family car for an evening out was tough. The first real time I was allowed to use the family Studebaker was the night of my high school graduation. True to form, as history would prove from time to time, I would often take a "golden egg" and turn it into a "goose egg". Later that night, after graduation, a bunch of us hopped in my Dad's Studebaker and headed to the beach for a night of teenage "action". Well, with a few beers in me and minimal driving experience behind me, I demolished my father's call running into a telephone pole in southern, Maryland racing others to the beach. Well, I woke up the next morning with a splitting headache and concussion in the Annapolis Hospital with my father hovering over my bedside with tears in his eye. Why, I thought he was crying because I had been hurt, but no, he was devastated because I had destroyed the family car and he had no car insurance!

So much for parental concern. Further to add insult to injury, the State of Maryland sent me a bill for $25 for destroying the telephone pole. Needless to say, I was grounded and didn't drive a car again until I could afford my own. As it turned out over the years I did not have the most outstanding record for driving automobiles. During past years I have collected some 15 accidents and totaled or demolished 4, including my father's Studebaker, two Volkswagen beetles, and my old pride and joy a beautiful Sunbeam Alpine sports convertible, plus three others, in which I racked up a couple of head injury concussions, a fractured skull and one serious broken arm that required two major surgeries. The insurance companies hated me! It wasn't that. I wasn't a good or competent driver, it just seems that I was a daydreamer and always preoccupied with thoughts on more important matters than driving. A bizarre remembrance, later on, in my twenties, when I worked for Remington Rand Univac, I was assigned as the personal chauffeur to drive General Leslie Groves, the Director of the Manhattan Atomic Bomb Project, and later a Director of Univac, to different military locations around the D. C. Area. He would always ask for me to drive him when he came to town, and luckily no one ever told him about my outstanding driving record. More on General Groves later!

Oh, almost forgot! Before I move on to the college years, I promised a little bit more on the "virginity awakening". Back a few years when I was a paper boy delivering the Daily News, one of my customers was a young woman around 21 years, who had moved from Ohio to take a job working for the Federal Government. I had been delivering the paper to her for a few months, collecting her bill in the evening each week after school. She was very nice, polite in every way, and seemed to enjoy seeing me whenever I stopped by to collect for the paper. And then it happened! It was winter and cold, and when I knocked to pick up her bill, she appeared at her apartment door in a sexy nightgown and invited me in for a hot chocolate or hot cider (I don't remember which I was so flustered). The hot chocolate led to the sofa; the sofa led to the bedroom and the rest I'll leave to your imagination. Needless to say, at 14 years old, I suddenly felt 21. A few months later she moved away; no fair wells, no notes, no conversations, no more hot chocolate or hot cider, who the hell remembers, she just disappeared from my life and left me no longer pure of heart, soul or body. No longer celibate!

Chapter 6 – University years

That summer after graduating from high school, I started working full time with the U.S. Post Office as a letter carrier in my surrounding neighborhood. A summer job with the Post Office was considered a "plum" by any kid that would luckily get one of the summer substitute routes. I was fortunate in having a neighborhood friend, Dick Everett, whose mother worked for the main Post Office in the personnel department, and through her I worked all summer "carrying the mail"' including magazines like the Saturday Evening Post and Life, which I swear weighed over five pounds each. My skinny shoulders, carrying the mailbag pouch, again looked like worn out craters by September. Canny old timer postmen always seemed to know what days these magazines would come and need to be delivered, and made sure they took vacation or sick days then, leaving it up to us, the substitute carriers to "carry the load". I needed the job because I had been accepted to the Catholic University School of Architecture in September and knew my parents would not be able to pay for the tuition. Wasn't sure myself if I would be able to save enough to go since back then student loans were did not exist. Besides, the backup plan was to go to the local public college if I couldn't raise enough for Catholic by September, and in that case I would need a car, since the college was quite a distance away in Northwest Washington.

Though I worked all that summer, I was able to get away on weekends and join some of my friends on several trips to "Wildwood by the Sea" in New Jersey, just a few miles from Cape May. Wildwood was crazy, lots of swinging bands, great parties and a great meeting place for boy-girl. And lucky me, again I met the lady of my dreams, Lillian Poleo, from South Philly, literally under the Boardwalk, where she was trying to get some shade. 'No, the Platter's song was not created for us, but it certainly fit the scene for the two of us. As it turned out every weekend for the rest of the summer I found a way to get to Wildwood to see her and before the end of the summer, she had my high school ring! We were engaged! - Didn't last long though. She setup a dinner at her home to "meet the folks" in Philadelphia near the end of the summer. I borrowed a friend's car, drove to South Philly, the heart of "Mafia" land and spent the Sunday afternoon being interviewed by her parents. Looking back, I just knew that her father had to be related to the Gotti family. It was quite obvious to me that “ole” man Poleo did not have me in mind for his daughter. So another breakup and another broken heart! My high school ring came back to me in the mail, a couple of weeks later!

Concentrating on school now became the priority. By then, it became clear that I didn't have the funds to pay for tuition at Catholic University, so plans changed and I enrolled as a freshman at Wilson Teachers College at 14th and Harvard Street in the City. What a comedown, there would be no architectural school now in my plans, maybe never. Tuition at Wilson Teachers was $75 a semester, a sum I could afford, and enabled me to use my savings to buy a car that I needed to get to school. Luckily, I had a friend, Warren Hosier, who lived nearby and was selling his 1941 Dodge for $500. So I bought it! My first car! In my mind it was a Rolls Royce. It had automatic drive and no "running boards".

Wilson Teachers was certainly not "Harvard on the Potomac". It was a small inner city college setup to educate students for a lifetime teaching career, and many of my friends that went there did just that and most that did, taught in the D.C. Public school system, many later becoming principals or high level government officials in the D.C. school system. Since schools were still integrated at the time (1949), Wilson was an all-white college whereas the minority teachers college, Minor Teachers College, was located just a few blocks away from Wilson, which all black students attended. In 1955, with integration ending, both Colleges were merged and became known as the University of The District of Columbia, commonly now known as UDC. The college was not large, really just slightly larger than my high school and had a total school population of probably no more than 300 students. It truly was an inner city school and serviced mostly kids from local public high schools, Coolidge, Roosevelt, McKinley Tech, Eastern, Wilson, Central and a smattering of those from various private and catholic girls and boys high schools.

School started that fall with hardly a hitch and I fell into the normal routine of school. Since the school was small it did not take long to meet just about everyone in my freshman class, as well as many of the upperclassmen. Academically the school was excellent, and the teaching staff outstanding. Having attended larger Universities later, there was something special about Wilson. Because of its size, there was an aura of closeness between the students and teaching staff that seemed to grow, as the months rolled by. Most of the students were from middle-class Washington families across the City, just as they were in High School. The only difference now was that the student body consisted of a cross section of Catholic, Jewish and Protestant kids, and the teachers did not wear catholic habits! It at first seemed strange to me to be taught by female professors in dresses, with no head coverings, and professors in coat and tie without a roman collar, but I soon got used to it and started to wonder why the Church made nuns wear that heavy garb, especially in the heat of the summer months. Lucky for them, church law changed in later years and they were allowed to dress like "normal" people. So we each set about getting to know each other and socializing together, both at school and on weekends. Remember, Wilson was a commuter school, so everyone went home each day and weekends, but we found ways to meet at parties, joint dates, basketball, baseball, or track events. I joined the College track team and ran in low hurdle and mile events, (note, Jesse Owens would never ever have had to worry about me), competing against other Teacher Colleges in the region. Schools like Towson Teachers in Baltimore; Frostburg Teachers in Frostburg, Md.; Pennsylvania Teachers College just outside of Philadelphia, and an occasional meet or game outside of our league, such as Gaudette College School for the deaf in Washington, which was and still is a College for the deaf. In two years of running track, I only won one race, and that happened against Gaudette. I think maybe it might have had something to do with my "hearing" the starting gun first, I mentioned above that it didn't take long for any of us to meet and pair up and that happened to me as well. I was taking a freshman English class, sitting next to a seventeen year old Irish/German gal with amazing blue eyes that lived in N.W. Washington and had just graduated from Coolidge High School that spring and wanted to be a school teacher. So, we hit it off together and started dating on weekends. I got to meet her parents early in the game and it was like revisiting the "Gotti" family in Philadelphia again. Oh, by the way her name was Pat Kreiter; she was Episcopalian, as was her father, and devoted to their religious beliefs. I'll jump ahead for a second and let you know that I eventually married her, but the intervening years!!! I'll talk about them in a little bit too, but what probably kept us going was her mother. Whereas her father was for a long time not so friendly, her mother, who was born Irish and catholic in Foggy Bottom, D.C., adjacent to Georgetown, took a liking to me. Mrs. Kreiter, maiden name Callan, was a slender brunette with blue eyes and freckled face, and looked as though she had just stepped off the boat from Ireland. She would spend time talking with me about her life growing up in Foggy Bottom and over the years I listened to and heard an awful lot of interesting stuff. Stuff like the fact she lived next door to Kate Smith and was a childhood friend of Kate's. You all know Of course who Kate Smith was! If you don't, well Kate was the singer who sang “God Bless America" at major events and made the song popular and became one of America's favorite idols, along with many others at the time, like W.C. Fields, Mae West and Rudy Valle. Which reminds me, my Dad, grew up on South Carolina Street in Southeast D.C.? In the same block neighborhood as John Phillip Sousa, the famous marching band leader of that era. So, I had something to share and brag about too! Now I’m not going to ask you if you remember John Phillip Sousa! If you don’t, Google! Mrs. Kreiter also knew and grew up with a guy by the name of Marty Gallagher, an Irish tough who grew up in Foggy Bottom and fought Joe Louis in a non-heavyweight fight. Gallagher at The time was being touted as the new "white" heavyweight fighter on the scene. Now Mr. Kreiter, he was something else! James Monroe Kreiter, they called him Monroe, was as you can guess, named after President Monroe and was a piece of work as the old saying goes. A short, diminutive, bald headed, cigar smoking, gruff speaking, bigoted towards Catholics, Jews, blacks type of guy, who was impossible to reason with. Kind of ironic that in the future, his daughter would marry a devout catholic and his son Jim would eventually marry his second wife, Deborah Ennis, a Jewish woman from Baltimore. Monroe was a Government worker, a printer at the Government Printing Office, who walked several blocks from home to the streetcar line that then took him to work every day into D.C. And then back home at night. Or most nights! Monroe had a thing for drink and conversation, and his favorite "bar stop" was conveniently right where the streetcar dropped him at night, so many a dinner was served without him I'm told. Dating his daughter, Pat, was not the easiest thing to do with him around all the time. Sometimes I prayed for the bar to hold him over night on weekends. Pat had a curfew of 10 o'clock on a weeknight and 12 on a weekend. Tough standards to meet if you were at a rollicking party! We used to roller skate occasionally at one of the local rinks and he would not allow her to wear short, sexy, skate skirts, so she would have to smuggle them out of the house to wear at the rink. He was an all present, omnipotent god" in charge "chaperone. He or his wife never went to bed until I dropped her back at the house after a night out. If we dallied too long in my car after a night out, in front of her house, there would always be a tap on the car window from one of them indicating it was time for me to go home. Things did change though over a period of a couple of years. I finally found out how to win! The old guy had been a corporal in World War One and served overseas during the war, but not on the front lines. He was a supply clerk miles from the front! But He won the war and loved to tell you how. So one night I was invited to dinner, just the four, and after dinner WW1 commenced; the women disappeared and the replay of the war began, using salt & pepper shakers, spoons, forks, glasses, anything to portray the biggest battle of the war that he was involved in winning. Bleary-eyed and exhausted after several hours later, I had gained a new pal. From that night on, he forgot that I was Catholic!

Pat had one older brother, Jim, who lived at home, joined the Navy and graduated from the University of Maryland business school. Jim had one of the most outgoing pleasant personalities of anyone I have ever met. He was marketing oriented, could engage in conversation with anyone he met and so it was natural that his career would move to working in the marketing and sales world. So after graduating from Maryland, he went to work for the Marriott Corporation, when it was substantially smaller than it is today. Over those years that he worked at Marriott, he met and worked directly with members of the Marriott family and knew Bill Marriott on a personal basis. After a stint with Marriott, he joined up with Chestnut Farms Dairy in a marketing role and joined their "Sealtest" ice cream division and spent the remaining part of his career working for them, developing and controlling major retail accounts with large food chains stores such as Safeway, and Giant, as well as multi Drug Stores chains. Like I said, earlier, he had, and still to this day, has one of the most engaging, pleasing personalities of anyone I have ever met. In all the years I have known him; I can never recall his ever saying an unkind word or backhanded comment about anyone, except maybe a Republican! Jim is a "died in the wool" Democrat and you couldn't get rid of those liberal color stripes using the strongest dosage of bleach. As he was out working in the business world, Pat and I were still at Wilson, struggling through the ups and downs of classes and our personal relationships. One day up, one day down. Most of it to do with religion- She was very firm in her beliefs and wasn't convinced that settling down with a Catholic was a good idea, since In those days if you married into the Catholic faith you had to agree to raise your kids in that faith, and she wasn't ready for that. Anyway, while all that was taking place, she introduced her brother to one of her sorority sisters and classmates, Barbara Butz. Barbara was a very bright and attractive woman majoring in math, along with me at Wilson and took an immediate interest in Jim. They married later after school graduation; moved to Rockville, MD, for a number of years where she taught High School math in one of the local public high schools. But they never had children of their own and with the passing of the years, I can only guess that the marriage dulled, like so many do these days and Barbara ended up having an affair with one of her fellow teachers and so, the marriage ended. The last we heard, Barbara and the teacher married, moved to Chicago, and raised four children of their own.

Meanwhile, back at my home, still on Franklin Street, my sister had grown up quickly under my own eyes. She had passed through 8 years of school at St. Anthony's elementary, then on to High School at Holy Notre Dame Academy, a catholic girl's school in the City, graduated and shortly thereafter married her high school sweetheart, Charlie Lanygher, who had just graduated from Gonzaga High. They settled down to a new life and ultimately four kids. Charlie went to work for the Government and my sister awaiting the birth of their first child Larry, ultimately ending up with four; two girls and two boys. Sometime later as we pass through these pages, I'll tell you more about them, and also a little on the life of my brother Phillip. Also, during this time, my stepmother who had been childless well into her forties, suddenly became pregnant and conceived her first, a boy they named Michael.

It soon became apparent to me that the money that I had saved the past summer working for the Post Office was beginning to run out. The cost of running my car, college expenses, dates and other entertainment was beginning to take its toll, so I had to find a part time job after school. Finding part time employment back in those days was no better than in today's world with 9 percent unemployment. But again the lucky gods showered me with a part time job at the D.C. Public Library, across the street from the old Arena Stage (Hippadrome in the Round), and I began working there in the evenings and Saturday, stacking books on shelves and checking them in and out for customers. It was there, at the checkout booth, that I met a young attractive woman, Yolanda Dewhurst, from upper Northwest, who was going to school at Georgetown University, who was friendly, outgoing and fun to be with. You see, about that time Pat and I were having usual teenager issues and had sort of broken up for the time being, so Yolanda entered the picture for a while and we became a heavy "twosome". Yolanda lived on Veasey Street in an upscale part of Northwest Washington born of wealthy parents, (her father was Undersecretary of some Federal Agency), but she didn't seem to care that I had nothing but a broken down old Dodge and a “crappy” part time job as my only assets. But her parents cared! And it wasn't much later, when I called her at home, that her mother answered the phone and informed me in a nice polite way that it would be best if I forgot about gorgeous Yolanda and went on with my life. Well, my interpretation of that call was "Jack, you're a nice guy, but you're driving a junky old car; you don't have any money; you're going to a two-bit college, going to grow up to be a low paid teacher, and I have better plans for my daughter". So, hat in hand, I called Pat one day and asked if she would like to go see a movie!

College months rolled by, the part time job was beginning to replenish my bank account, Pat and I were still dating, getting more serious and our circle of friends kept expanding. At school, we both were close friends with several other couples, Cliff Kendall and Camille Lehman, and Wally Mitchell coupled with Jean Mitchell, and Jim Lamont and Lena Bartholomew all whom ultimately married each other. Summer came and Pat and Jean found a job for the summer in Wildwood waiting tables in a restaurant, and I continued working at the library. Weekends were spent traveling back and forth to Wildwood to see each other and take in all the best of the town, watching and listening to the original "Hubcaps", the Inkspots playing at the Rainbow Bar in Wildwood Crest, Tony Pastor, a popular singer, at the old Lincoln Hotel and, still to this day performing, the great Chubby Checker who I watched sing and dance as late as 2010 in Fenwick Island, Delaware, near my summer home in Rehoboth.

Being 17 almost 18 is a fantastic time for kids. You're just one step out of the cradle, but yet you feel grown up, with not a care or worry in the world! Life is marvelous; fun, exciting, new every moment, always changing quickly; most of the time for the better. But then all summers end, and before you knew it our sophomore year at Wilson started. It was obvious to me that Pat really did want to continue on at Wilson and graduate to teach elementary school. I really didn't! I still wanted to be an architect and was taking as many courses in math and art that I could, hoping they'd be transferred to Catholic U. in my junior year and I would find a way to pay the tuition bill. It was at Wilson that I took several art classes with Howard Mehring who became a friend of mine, and who by the way dated Pat when we were in our breakup phase. Howard was a gifted painter in College, but his best days were in front of him. Howard went on to be one of the six nationally acclaimed artists that founded the Washington School of color, an artistic movement that started in the late fifties through the mid-sixties. The group consisted of Gene Davis, Thomas Downing, Ken Noland, Paul Reed, Morris Louis and Howard. It was a school of visual art movement where artists painted largely abstract works using stripes, washes and fields of single colors of paint on canvass and became a popular art movement during that period of time. Sadly, Howard became an alcoholic and died in his early thirties, and unfortunately because of his early death, he left behind few paintings to remember him by. But with a little bit of luck I was able to locate three of his contemporary works which are now hanging in my Virginia home. The Color School era was extremely popular at the time, and many wealthy collectors bought and collected their works and major art galleries and art critics extolled the quality, brightness, simplicity and sheer magnificence of their work that they created. Gene Davis was a favorite of mine, and to study and view the brilliant works of varied stripes in his paintings is a joy. The man was a genius and one of the most gifted artists of his time. From time to time, there comes resurgence and interest in their works, and just recently in 2011, the Corcoran Art Gallery in Washington brought together a number of these paintings from that era in a major show.

By now we were" seasoned "college students into the second year, becoming more serious about the future and trying to become more focused on what we wanted to do and study. Pat had decided to concentrate on majoring in elementary education whereas I continued taking courses I could transfer to Catholic, i.e. math, drafting and art classes. She had joined a sorority, and taken a part time job at Goldenberg's Department Store, a Washington retail institution back in the 40s' and 50s', not far from where I still worked at the D.C. Library, So we were able to see more and more of each other, both at school and after work. Her brother was still dating Barbara and we would oftentimes double date with them. Also, there had been a significant new development in the Nation. Earlier, in June of that year (1950) we entered into a war with North Korea and several of our classmates left school to join the Armed Services. War had broken out between North and South Korea and since we were a protector of the South, we were brought into the conflict that turned out to be a vicious and difficult time and when it ended in 1955 in a shaky treaty, the rest of us had finished College and were moving into career jobs, or graduate school programs. Years later, after graduation from school, one of our friends, Duffy Hutton, who graduated from Wilson joined the Air Force and fought in Vietnam as a jet fighter. He was shot down in Vietnam and was imprisoned for close to ten years in Hanoi, one of the longest imprisonments of the war. He was there with Senator McCain and others during that time. After his release, we all celebrated his release when he returned to D.C. with his wife, a nurse he met while recuperating in California when he returned. It was amazing to see and talk with him-he had aged but did not display or talk about any of the years spent there and the torture and pain he had to have suffered. One of my best friends, Howard Thomas, a few years older than myself, had the misfortune of being drafted late in World War II, and later separated from the Service; then was recalled to serve in Korea, and was sent to the front lines at the 36th parallel, where again he was seriously wounded on two separate occasions and separated again. For these two wounds he suffered in Korea, he was awarded two purple hearts, and a head full of bad battlefield memories. Two wars, two typical tragedies!

Things were changing at home as well. My parents had decided to move to Silver Spring, Maryland that summer of 1951, and since I was still living at home, I moved as well. My sister by then had married, was out of the house, tending to her first born with her husband Charlie. Phillip, My brother, graduated from St Anthony's and with no interest in College at the time, he decided to join the U.S. Navy.

Sophomore year ended, summer of 1951 arrived and I still continued working at the Library. Pat had decided to take a part time job in Ocean City, Maryland and by then I realized that architectural school was a dream that wasn't going to happen. So, it was then I decided to change direction career wise and with several student friends from Wilson, Cliff Kendall, Bob Zamsky and a boyhood friend from my old neighborhood, Ray Lunceford, we all planned to enroll in the business school at the University of Maryland that fall. Changing school direction cost me almost a year of lost college credits, so I basically had to start back at a sophomore level since Maryland would not accept them. Enrolling at Maryland was like moving to a different planet. From a school of 300 or so students in one small building to a campus like Maryland with tens of thousands of students was an overwhelming experience. From classrooms of ten to those with over fifty, and sometimes larger was jaw dropping. Campus buildings and dormitories, football and basketball plants stretched for acres engulfing miles of campus area. The move to Maryland was like the move and mood of the Nation at the time. We seemed to be escaping the dark and depressive spirit of the late thirties and forties and moving into a more relaxed and expansive world of excitement and newness. Young adults seemed to be unshackling the depression mindset of their parents and experimenting with new ideas, expanding their desires in terms of the education and future lives they wanted. We didn't want to be like our parents! We wanted to learn more, travel more, have more, spend more and enlarge the number of friends and acquaintances we had in terms of backgrounds, varieties of interests, thereby growing ourselves, In other words to get out of the limited boundaries of our personal neighborhood! Maryland, in a way, offered a start in that direction. For the first time in my life, I began to meet and interrelate with a broad section of new friends and acquaintances from around the country that had enrolled at Maryland.

Because I had now moved to Maryland, and considered an in-state resident, I was able to handle the cost of tuition and other expenses myself by continuing to work a part time job after school. My old Dodge by now had begun to fall apart and all of a sudden, here comes another expense, another car, that I did not need, but transportation was a necessity, since I was still living at home commuting both to school and to work. Well, now comes an embarrassing story as to how I got my second car through the help of my Uncle Eddie, my father's younger brother! Uncle Eddie was the "black sheep "of the Ballenger family. Or, I should say one of the black sheep in our family, but Eddie stood out amongst them all. He never finished high school, constantly in bar fights, talented with his fists, never really amounted to much in life, worked menial jobs, and had been married five times, to believe it or not, women who were lovely women. He was a good looking dude with an engaging and persuasive personality that resonated with people, especially women! He liked me personally and deep down I think he was impressed with the fact that I was the first Ballenger to go to College and wanted to help with the car thing. At the time, he was holding down a job with one of the major laundry companies as a driver picking up and delivering laundry from different customers in the City. Back then many people still did not have equipment in their homes to wash and dry their clothes, or if they did they still sent things out to be laundered or dry cleaned. Next thing I know Uncle Eddie calls me and tells me he has found a car and arranges for us to go look at it at a private home in Northwest Washington. So away we go two guys in a company laundry truck to look at, soon to be, hopefully my second Rolls Royce. We get to the home, park the truck and Eddie knocks on the door. Well you could have knocked me over with a soft drink straw! Standing in the doorway is a beautiful woman, a striking beauty, and as it turns out, all alone; husband at work, no children at home. We are invited in, my Uncle immediately has his hands groped around her waist supplying her with a little mouth to mouth resuscitation and I'm standing in the hallway dumfounded. A few seconds later, or maybe it was minutes, he introduces me to her; we go into the kitchen for coffee; he mentions the car; she goes and gets the keys; he gives me the keys and tells me to go for a test drive and come back to the house in about an hour. Which I do! The car I love! Turns out it's a 1951 Henry J, a car made by Kaiser Aluminum, a company that also made the now extinct Nash line of cars, For you car aficionados, the old Nash Ambassador, the Metropolitan, and lastly the Henry J, named after the president and founder of the company, Henry J Kaiser were automobiles built by Henry after the War to compete with the big 3 manufacturers at the time they were beginning to build cars again instead of tanks and jeeps. Why am I telling you all of this, probably just trying to waste some "word test driving time" before I had to go back to the house. The car was perfect and just what I needed for transportation, and to me almost brand new. The concern now was, I only had $1200 of cash in my pocket and worried that it would not be enough. Turns out that was not a problem at all. When I get back to the house, I knocked and when there was no answer, I walked into the hallway and sat. Waited for about a half hour and there they come, down the steps in a glorious, happy state of mind. Well Uncle Eddie asked me how I liked the car; I told him it was perfect and before I could ask her how much she wanted for it, Eddie pipes up and says "give her $900"! I was flabbergasted and was shaking when I pulled out of my pocket and counted $900 for the car, and handed it to her. With that he turns to me and says "ok kid, take the keys and get moving, I'll take care of the paperwork- see you later"! Driving back home that day, I was thinking, a laundry truck driver job was something I should be considering. The hell with college!

But return to college I did and entered the School of Business, changing career dreams altogether. So my schooling changed from how to design and create beautiful buildings to an education that taught me how to market, sell or lease beautiful buildings! I soon found that I liked the business and economic world of study, and even then began to think of one day starting and owning my own business. At least I was beginning to up my sights and think a little differently from the time my old boss, Ms. Sigman, at the Kann's department store heard me say that one day I hoped to make $200 a week! In addition to schooling, Maryland U. had developed a nationally ranked athletic program in major sports, primarily in football. Paul (Bear) Bryant had over a few years created a formidable football power by the early 50s', but decided to return to coach at his alma mater, the University of Alabama. Replacing him was Jim Tatum, who continued taking the teams he coached to major Bowls, including two Sugar Bowl and one Orange Bowl appearances, while I was there. On the Orange Bowl occasion I scraped together enough money to take Pat to the game in Miami over the initial objection of her parents. Remember this was the early 50s' and parents believed in celibacy for their daughters forever! We got around the problem by inviting her brother Jim and Barbara, who he was still dating, and all four of us with Jim driving, took off for Miami and spent the week soaking up the sun and excitement of winning the big game, getting sunburned and reveling in the fantasy of what great jet setters we had become. Uh, by the way, It was this trip where Pat, I think, decided that maybe the catholic marriage issue could be worked out between us and from then on things got a little bit more serious between us (for a while!). Back to football, Bryan and Tatum have been recognized as two of the best college coaches ever in collegiate football halls and in that era much emphasis was placed on sports activities versus academics. People like Tatum and" "Bud" McMillan, the basketball coach, later replaced by "Lefty" Driesell" became more recognized and powerful than the academic community at the University, so academic standards were often reduced to allow some of the best athletes in Maryland history to attend the school. Football athletes like, Bobby Ward who became the University first consensus All American player leading the team to 10 - 0 record in 1951; and Jack Scarbath, the school's first All American quarterback ; Stan Jones, Gary Collins, all who became standout Pro players with Stan Jones landing in the NFL Hall of Fame. Major funds were set aside from school budgets to develop the athletic programs to a national stature, and major it did become. Many of these kids came from the mining towns in upstate Pennsylvania, which became a central recruiting area for Tatum, as it was for Joe Paterno at Penn State and they were "big"! I remember especially two brothers, the Modileskis', who went by the nicknames Big Mo and Little Mo, who were in several of my classes. When they walked together down and thru the campus on their way to class, you could spot them a mile away, a foot taller than most and shoulders wider than the football field they played on. "Little Mo" was a nice guy who took a statistical analysis class with me (his second try) and his heart was not into it. He was there to play football not to take classes. As I recollect, he hardly ever showed up for class, but I'm sure he passed, and he did graduate thanks to Mr. Tatum and probably became the mayor of his home town today after making a fortune playing pro football.

Meanwhile, I had scheduled my classes so that I could work extra hours, some afternoons and evenings at the D.C. Library in order to keep the personal bank account alive. The added costs of school and expanding social activities were "killing me". My expenses such as gas for the car went from 17 cents a gallon up to 25 cent and were heading higher. Don't laugh, back then that was a lot of money for a gallon of petrol. So, I had to find other work to make more money and still handle school schedules. Again, thanks to good old "uncle Eddie"', or I should say his wife Marilyn, his fifth and last, I was able to get an evening job working for the Republican Minority Clerk, Tom Langford, in the Rayburn Senate Office Building, My Aunt (5th) who was the top administrative secretary to a congressman from Missouri was a savior, the pay was outstanding and the job interesting. Our office and operation did all the printing work for Republican Congressman on the Hill, except for typing. In the early fifties there were no personal or office computers, so all printing had to be done in old fashion print shops. That's what we did! Langford was just a young man himself, in his early twenties from Indiana, a young Republican, who performed well for the party and for it he was given this patronage job. He was a quiet guy, but easy to work for and I did, until I graduated from Maryland. Even remember, he lent me his "Elvis Presley" red Cadillac Convertible to take Pat to my senior prom, and gave me $100 for spending money as a graduation gift. Years later, when Pat and I moved to Potomac, Md., Tom also moved there, and we socialized a number of times with him and his wife, Pat, who was an outstanding recognized tennis player at the Congressional Country Club in Potomac and won a number of single championships around the City during her playing career. Unfortunately, she died of cancer, and Tom heartbroken moved to a small room apartment at the Club and began to lose interest in life and connection with people. He was a good friend and we kept in touch for many years but lost contact with him after his wife died, when he became a shy and reclusive person. By coincidence, I ran into him at Laurel Racetrack in 2010, when I was running one of my thoroughbreds, and invited him to sit with my party, since he was by himself. He kindly thanked me, but refused. So shy, so sad! It was tragic to see someone you once knew, full of life and ambition retreat into a cloistered, lonely life.

Since Maryland was a State University, there was an Air Force ROTC program for all male students which all of us joined, including all of my friends that had transferred from Wilson Teachers. It was a general military program for the first two years and then you could continue Officer training for your remaining college years and expect a commission in the Air Force once you graduated. However, for part of the program one would have to do "field duty" for one month in the summer. So, Ray Lunceford and I were sent to Albany Air force Base in Albany, Georgia to do our active duty, while our other friends were assigned to other Air Force bases.

We were quite excited about a possible after school career in the Air Force, and settled in our BOQ, dreaming about flight school; getting wings pinned on our uniforms and flying jets to help win the next big war! As you can probably guess, it didn't turn out that way. We were both there for two weeks before we were called in for a meeting with our ROTC commanding officer and told that we were being removed from the program for medical reasons. It turned out that my asthma medical records would not pass muster and Ray was turned down because of a mysterious kidney problem. So after a few minutes of condolences and "good luck" handshakes and given a few dollars to get back home, we left the base and decided to hitch hike home. Now that became the trip back of the century! Ray and I had decided to save the Air-Force money and hitch hike back to Maryland, so hitching different rides from town to town, state to state, we hooked up with a driver in a beat up old Chevrolet, who was heading back to South Carolina from his job in Georgia. He stopped and asked "where are you headed"? We told him North to Maryland, so he nodded; motioned us to the back seat and off we went speeding down the highway at close to 100mph, when we, in a petrified state, noticed that he had a bloody rag tied around his right hand, which constantly moved from the steering wheel to the passenger front seat, picking up a bottle of whisky and taking several occasional swigs as he roared down the road. Then he started telling us how he had just cut off his right thumb at work and had just decided to leave the job and head home, incidentally, with the severed thumb in a dirty bag on the front seat. It obviously hurt like hell; he was babbling in an incoherent way, drinking like crazy and just steps away from killing us all as he sped on! It took us about twenty miles for us to finally convince him to let us out, telling him "we made a mistake, we really live in Atlanta and were going the wrong way". I can't believe he made it home alive. They probably buried him somewhere on the outskirts of Aiken after the crash, hopefully with his severed thumb. Whew, what an experience; after that we interviewed each driver that picked us up before we got in a car. Only kidding, but if a Greyhound bus had come by first, I would have grabbed it in a heartbeat!

It was now June 1953 and graduation time for Pat and other friends I had left behind at Wilson. The two of us were still dating, but since I had lost a year's worth of college credits transferring to Maryland, I still had one year to go, so we decided to put off any serious marriage talk for a while yet. That summer, after graduation, she spent the summer in Ocean City, Maryland with several of her sorority sisters that had also graduated with her, and I, still working on the Hill during the summer, spent my weekends in Ocean City with her, before she came back to Washington to begin her teaching career in the fall. Ocean City was a much smaller mirror of Wildwood, not as much Doo-Wop, smaller, lesser known bands and singers, bars versus night clubs, but as Wildwood catered more to the youth of North Jersey and Philly, Ocean City, which was closer to D.C., attracted more kids from D.C. and Maryland. Besides, by now we were following our girlfriends, not searching for new beaches, so Ocean City was the lighthouse beacon to follow for the summer of 1953. We were still hanging out with the same group of couples that had become close friends, Cliff Kendall and Camille Lehmann, Ray and Ann Lunceford, who had recently married, Wally Mitchell and Jean, and from time to time, Jim, from Wilson, with his girlfriend Lena, whom he later married (me as best man) and also Pat's brother Jim, and his new wife, Barbara.

Ocean City at that time extended from the Inlet to 15th Street. To the north was Worcester County, which was pine groves, marshy meadows, and sand dunes. Along it were a handful of private summer cottages and rooming houses, owned by mostly families that summered in Ocean City from nearby metropolitan areas. Others rented the same apartments year after year, or in Pat's case, she and her friends would rent rooms in one of the rooming houses near the beach area, and eat their meals where they worked. Everyone looked forward to summer, and it was always a joy when it arrived, after a long winter. Hotel owners, and boardwalk shopkeepers, looked anxiously toward a profitable season and those students, like Pat, were hoping to find a job, which she did as a waitress in a small restaurant, extolling her past waitressing experience in Wildwood to get her job. A teacher's degree didn't help! Older students found work on the fishing boats of the marlin fleet or working as waiters in the smaller restaurants or carryout or renting umbrellas and beach chairs to visitors on the beach. Larger hotels and restaurants usually hired adults or had their own staffs that returned each year and lived in small rooms under the hotel. It was a carefree time with much laughter and there was always a major league baseball game being listened to on an old fashion portable radio. The Baltimore Morning Sun newspaper headlines everyday showed positions and battles of our troops in Korea and along with radio, this was primarily how news was reported, since few families had television at the time; and the news was not usually good, military reports in general were glum throughout the summer.

On some summer evenings we would walk down to the Pier Ballroom, which was a facility provided by the City for young people as a gathering and meeting place. It was open to locals and visitors alike for a 50 cent fee which jumped to 75 cents on weekends, when there was live music. Jitter bugging, the Mexican hat dance and the bunny hop were very popular, and usually performed to the live music of the Charlie Shockley orchestra from Salisbury.

The City has changed today, more crowds, buildings, and traffic congestion. The rise of so many high tower condo buildings and hotels stretching across the Ocean City beach line have blocked out the beautiful ocean views of so long ago, except for a few of the more fortunate that own a beachfront condo. The building and completion of the Chesapeake Bay Bridge in 1952 created a major increase in people visiting the beach areas and the City began it's fast paced journey to becoming Maryland's second largest city.

Today, the dunes and bayberry bushes are nearly gone. The wild beauty and the City's unique small time charm lost forever!

Summer of 1953 ended and everyone headed back across the new Bay Bridge, with more serious thoughts on their minds; Pat with starting her career teaching 4th grade students in the City public school system, and myself starting back for my senior year at Maryland with no thoughts of going back to the architect dream. I was tired of school with the extra year added on to get my degree, so the motivation was fierce to get back and finish and then go wherever that degree would take me. Pat was assigned to an elementary school in Anacostia, a part of the city in southeast D.C., quite a way from her home in Northwest, where she still lived with her parents. Luckily, I had taught her how to drive the previous year, so the first thing she did when she reached home from the beach was to go out and buy a new Ford sedan. So with a new set of "wheels!" and a full time paying job making $10,000 a year, she became even more independent!

I continued working on the Hill in the evenings and occasional weekends that last year of College and dating her whenever we could squeeze in the time. Ray Lunceford had married his high school sweetheart; Ann Meyers that year, Pat's brother was now married, Cliff Kendall and Camille, the same, and Jim Lamon and Lena had tied the wedding knot. The bachelor walls were crumbling! Further troubling, the war with Korea had in a sense halted with the Armistice in 1953, but the U.S. was still drafting men since there was a need for additional replacements for Korea and elsewhere. This need had been quite honestly one of the reasons so many college students had signed up for college ROTC Air or Navy programs around the country to escape the front lines of Korea. As college seniors we were well aware that we could be drafted after graduation, and in the case of Ray and me, we could very easily end up in the Army, even though we had flunked an earlier Air Force ROTC physical. No one wanted to spend their first year out of college freezing in a tent at the 38th Parallel in Korea. I'll tell you a funny story in a moment or two, but meanwhile my relationship with Pat began to drift again, as you can probably guess, seeing your friends around you awash in marital bliss, social conversations sort of ostracizing you when the talks centered around new careers, babies, their own homes and moving out from under their parents had its effect on us. I was still living at home with a young eight year old stepbrother, my sister by now had several children and my brother was out of the Navy, married with the first of his six brood. "I wouldn't grow up" as I was reminded every time it seemed I would get together with her. I think even the “ole” man, Mr. Kreiter, was beginning to think, "Well, now that she's picked a Catholic, when's the Catholic going to make his move"? By now he had retired from the Government and thinking of moving to St. Petersburg, Florida, which meant that Pat was now going to be faced with moving herself and they were concerned about leaving her in town by herself.

Graduation from Maryland came for me in the spring of 1954, with my parents, grandparents and Pat in attendance. I'm not going to bore you with a lot of minutia, graduations today are not a big thing, but in our family it was. I was the first Ballenger and/or Raymond to have set foot in a University or graduate from a college so, it was a big deal. I only bring this up because it reminds me of my father at that ceremony. You see, my father never got past the 6th grade in school, when he dropped out to become a baker apprentice at 11 or 12 years old and a full-fledged baker at 16 to help his family survive. I can remember him at the ceremony, mesmerized and awed by the crowds, the campus, and the tears of pride when he saw me get my diploma. The whole scene was unreal for him and perhaps for me too seeing him with moist eyes in the graduation crowd. In a way, that graduation seemed more important for what it gave him versus what it gave me. I don't think this scene was probably unique at all back in the 50's; this was really the first generation of American kids that had en mass gone to college, and I am sure there were tens of thousands American families proudly praising and thankful for their sons and daughters that had done the same.

Now, it became a matter of trying to escape the draft and maybe marriage for a little bit longer, and find full time employment in a new career. Well here I want to stop and tell you the funny story I alluded to above, funny, but in a sense not so funny for my friend Ray Lunceford. At the time the Army draft system was decided by lottery, and since our college draft exemption was now void because we had graduated, we knew there would be a good chance of having our numbers drawn before the end of the summer. Remember, Ray was now married and concerned about leaving his bride, while he froze to death in Korea. So I convinced him that we should both volunteer for the Army since it was sort of an unwritten Army policy that if you volunteer, you had an excellent chance of selecting your own MOS and getting a safer and more interesting assignment, maybe if you're lucky right in your "own backyard". After mulling it over, we both decided to enlist and put our names in as volunteers and were shortly informed to show up at the local recruitment office for interviews, tests and a general physical. It turned out to be an all-day affair; lots of recruits, lots of lines and since the lines were alphabetized, Ballenger was near the front and Lunceford somewhere in the middle. Midday, while standing in line we were given a box lunch to tide us over. Many in line were turned away for various reasons, a good number because of flunking the tests, others for medical reasons. My turn came and I was told that I was being turned down for medical reasons as once again my bouts with asthma surfaced and even the Army was not taking volunteers with what I guess they considered a major medical problem. Turned away, I left, and as I passed Ray in the line I said "hey, here's an extra lunch box, they're sending me home"! He was shocked, and six months later after boot camp, he was assigned to a "tent" in Korea, leaving his wife to fend for herself! Meanwhile, I had to sheepishly crawl home and figure how to give back all the going away presents given to me by Pat and friends. An embarrassing day! To this day, when we get together we always laugh when we talk about that time. Ray spent over a year in Korea in the supply lines of the Army and always likes to tell the stories of different provisions he had to give out to various enlisted guys, many who came from the backward countries of many poor state like West Virginia, Mississippi, or elsewhere, poor and unsophisticated were many. Hearing him tell the tale is hilarious; at the time in Korea there were still a lot of "camp followers", prostitutes that plied their trade around the Army bases, so Ray, in this "senior capacity" was also responsible for giving out prophylactics to soldiers in an attempt to keep the cases of venereal disease from spreading thru the camp. So as he tells the story, when a small young soldier would come into the supply room, kind of shy or have a difficult time asking for a pack, Ray would take a package from behind the counter and kiddingly ask him- "what size “do you want soldier, they come in small, medium and large? With a slight shuffle of the feet, the answer would always come back- "give me a Large"!

Chapter 7 – Hardware Years

"The Tin Can Years"

So the beginning of many years in the business world started in the summer of 1954 in Baltimore as a sales and marketing trainee with the American Can Company, a manufacturer of tin cans and milk cartons. The Company, commonly known as CANCO, was the largest company in the industry at the time, as well as the oldest, having been founded in 1901, from a collection of can companies around the country. It had a long and glorious history and was part of the Dow Jones Industrial Average from 1959 thru 1987. In 1986, an investment banking group led by Gerald Tsai, a well know investment banker and leveraged buyout genius acquired the Company, and renamed it Primerica turning it into a financial enterprise. In 1987, Tsai divested the can business by selling it off to Nelson Pelz for $570 million, who was acquiring other can companies including National Can Company, a major competitor of CANCO over the years. An interesting side note is that in 1987, Primerica was acquired by Sandy Weil along with his Commercial Credit Corporation which ultimately became known as Citigroup. Commercial Credit was a Baltimore consumer credit/insurance business which designed and built some of the early "charge plate" machines thru its subsidiary Central Charge Corporation, where credit card data was imprinted on a charge slip versus "swiping" as done today, and was owned by Control Data Corporation (CDC). CDC fell into financial trouble in 1987-88 and Weil was able to convince CDC to spin off the credit company and take control, leaving CDC with a remaining 20% in the company.

I was still living in Silver Spring with my parents which was a long drive to Baltimore each day, since Beltways and the Washington /Baltimore Parkway were just in the process of being built, so the main way to Baltimore was via the main artery north, Highway One which was slow and time consuming. I was still driving my old Henry J and used it for a while commuting, but I soon found a better way to commute, without moving to Baltimore, and that was taking the old B & O commuter train from the Silver Spring Station which dropped me in Baltimore at Cameron Station, now the site of the Oriole baseball stadium and the Baltimore Harbor; from there it was just a short walk to the office where I worked at Redwood and Light Street. I had decided to continue living with my parents rather than move to Baltimore, since all my social life and Pat were still there. Besides, Baltimore was still an old and dingy industrial town, just beginning to "rehab" itself, so Baltimore was the "work" site and Washington still the "play" site! I was hired by the Baltimore Vice President, Mr. Goodman, a middle aged executive who took a personal interest in me and while I worked there, he became like a second father to me. The starting pay was $400 a week, which back then was an impressive wage, and I felt like a millionaire starting my new career. Since the position was a two year training program, I was moved back and forth thru a number of the operations, including the marketing group, the advertising development entity and in the field, at the local canning production facility in Halethorpe, a suburb of Baltimore. I was assigned to an up and coming executive, Sid Barteau, who had recently himself transferred from the Company Headquarters in New York to the Eastern Regional Division in Baltimore. Sid, was young, still single, and at 27, just a few years older than me, but had been with CANCO for several years and was a tremendous help in teaching me the business. We became fast friends and many a night after work, spent time together, enjoying the better parts of Baltimore, including the Playboy Club next to the office. He had a beautiful baby blue 1952 Ford convertible which we drove around from customer to customer and before I left the Can Company in mid-1955, I became the owner of that car, finally retiring my old Henry J. With a steady job, and, an awesome new car, to impress Pat, her family, and my friends, life was starting out to be what I had hoped for.

Immersing myself into learning all I could, I quickly found out that I was especially attracted to the marketing and advertising parts of the business. The different stints at the production factory learning how cans and cartons were made was really not of interest to me, other than using what I learned in the factories to better help me in providing customers assistance when pitching products or solving problems. I enjoyed more the interaction with customers and looked forward to sales and marketing calls, first with Barteau and then later on by myself. Most of the customers I worked with were located in the Baltimore-Washington corridor and were some of the largest companies then, and many are still today. Companies like National Bohemian Beer, Pabst Blue Ribbon, Valley Forge Beer (my favorite-but long gone brew), Del Monte, who made cans for peas, string beans, fruits, etc. In addition, milk and juice cartons were sold to clients in that industry; companies like Chestnut Farms and Embassy Dairy in Washington and High's ice cream stores. I noticed quickly that the milk cartons we sold to various dairies for different milks, skim, regular, buttermilk, orange, grapefruit and others were all of a generic color, i.e. a cream like color, so that when stacked on counter shelves, it was difficult for customers to pick out milk from juices or other milks when trying to read the descriptive lettering on the carton. So I convinced a number of customers to do different designs for each carton and to better color code each product, for example, brownish carton for chocolate milk, white for white, orange or yellow for juices. Although this idea seems awfully mundane today, it caught on quickly around the country and most milk producers used this idea for years. Writing about cartons reminds me that when I joined CANCO, International Paper Co. controlled almost 90% of the carton industry. You may vaguely remember their "PurePak" carton which you opened by separating the triangular top seal, which never seemed to open properly. You'd usually end up cutting a hole in the top to get to your milk. CANCO had just designed and brought to market the "flip top" opener which was located at one corner of the top of the carton, which you flipped and then resealed. It quickly became popular and captured a high percentage of the business from International. My favorite customer in the year I worked there was a fellow by the name of Clyde Ehrhardt who was one of the buyers for products used by the High's Distribution network. Clyde was an easy going guy who was a center for the Washington Redskins when the legendary quarterback Sammy Baugh was still playing in 1948, at age 34. Clyde was from Georgia where he played his college football; and for that time, a big and gregarious guy at 6'1 and 230 pounds. He was still in his early thirties when I first met him, and we would occasionally get together for a drink or dinner after work, when he would mesmerize me with stories about his time with the

Skins and playing days with Baugh, Steve (Bugsy) Bagaras, Joe Tereshinski, Al Demaeo and other players of that era. He shared center duties with Demaeo, and loved to talk about what a privilege it had been to have played with Baugh shortly before Baugh retired.

With both of us now working full time during the week, Pat and I did not see much of each other except on weekends, so we seemed to be drifting away from each other the year I was in Baltimore. We were still steadily dating, but most of our friends had settled into a married life with babies beginning to spring up everywhere as well as in every conversation, so common interests with all seemed to be fading. By this time we had been dating for over five years and she was ready to settle down, get married, and to start a family, but even though I loved her, I just was not ready. I felt my career was still in an early stage and with a recent raise to $450 a week, there was still a concern that I would need to be making a lot more than that to really support a family, and the dream was still there to one day start and own my own business-which meant time and money. They may have not been the real reasons, though, I guess the years of catholic indoctrination just kept gnawing away at me; the words kept coming back "join the priesthood or marry your same kind"! We did spend time that summer going to the beach on weekends, other times enjoying dinners or theatre with friends like Frank and Mary Wright, Wally and Jean Mitchell, Jim Lamon and Lena, as well as others, but by the end of that summer we decided to start dating others. I joined the local Catholic 18 to 30 club to meet new people and soon started dating a striking red head from Ohio who had moved to The Silver Spring area. She was athletic, loved to ski and that winter we took a number of ski trips to nearby ski resorts in West Virginia and Pennsylvania with the club, meeting new friends and a few I had known back in high school years. One of my high school classmates, Jim Reis was a member, as was his sister Teresa, who I also dated a few times that year. It was here that I met Carl Peters, a scientist who worked at NASA Goddard Space Center, who became a close friend in those days and ended up as my best man at my wedding a couple of years later. As it turned out, I was offered a job with Remington Rand UNIVAC that summer and resigned from the Can Company. The era of the computer had just started five years ago and as it turned out I was there almost at the beginning of an industry that has revolutionized the world. Thinking back I could never in my wildest dreams believe what would happen to me in coming years and the people I would meet along the way.

"The Hardware Years" The Beginning

That summer I happened to run into an old friend of Pat's from Wilson Teachers, Elena Rispoli, who had graduated with Pat, but decided to work in private industry and had recently taken a job in 1954 with Remington Rand Univac, to learn and teach employees and customers how to program the first commercial computer called the UNIVAC 1. Elena informed me that the Company was expanding and looking for young recruits to join and become marketing and or system programmers. I interviewed with the Vice President of the Washington office, Jack Veale, and started working as a programmer trainee at the end of the summer of 1955, and although I was excited to begin working with computers, I was not enthralled with my starting salary. Veale had offered me a job beginning at $400 a week, which was like starting all over again at the Can Company but thankfully I had the foresight to understand that this new opportunity was going to be explosive and change the entire world's way of dealing with business and scientific data, and I could be a part of that. Remington Rand, as it turned out, was a company that manufactured a line of business equipment including typewriters, 90 column punch card calculators and bookkeeping machines, a new computer division and also a division manufacturing and selling electric shavers. Back in 1950, Jimmy Rand, the President of Remington Rand had purchased the Eckert Mauchly Computer Corporation (EMCC) from Drs. Eckert and Mauchly, the inventors of the first vacuum tube computer called ENIAC, which they built while working at the University of Pennsylvania's Moore School of Electrical engineering. Eckert and Mauchly left the School over a patent dispute and had obtained a contract with the Northrop Aviation Corp. to build a computer called the BINAC, which turned out to be a failure and not used by Northrop. Shortly afterwards they obtained a contract with the Census Bureau to build a computer for the nationwide census count that they named UNIVAC. The initial value of the contract was $400,000, but they vastly underestimated the value of the development effort, which ended at approximately $1,000,000. They then attempted to renegotiate the contract with the Government and when the Government refused to renegotiate the contract Dr. Eckert and Mauchly decided to continue with the project hoping that the publicity of the project would bring them additional business.

Among the first assignments given to Eniac, first all-electronics digital computer, was a knotty problem in nuclear physics. It produced the answer in two hours. One hundred engineers using conventional methods would have needed a year to solve the problem

Figure 4: The Eniac

As it turned out, at this time, the major backer of Eckert Mauchly Corporation, Harry Strauss was killed in a plane crash in October of 1949. With no way to further finance the Company they first approached Tom Watson, Sr. and Jr. In an attempt to sell them the Company, but they could not convince Tom, Sr. of the UNIVAC's viability. Watson, Jr. was later heard to say, "Dad did not trust magnetic tape, and felt that if you had a punch card you could see it and hold it, but with magnetic tape data was stored invisibly on a medium that could be erased and reused". So, EMCC went to Jimmy Rand, who purchased the Company. Rand's Company had earlier in 1949 built the first line of business computers called the UNIVAC 60 and UNIVAC 120 which were programmed by hand by means of an electronic removable panel plug board. It sort of looked like, and reminded one of a smaller version of a Lilly Tomlinson telephone switchboard. The very first user of the UNIVAC 60 and 120 computers was the Internal

Revenue Service which used a 90 column punch card for input and output. These systems were a slow but small step up from the punch card days, but Rand had sensed that customers were going to be looking for faster and larger capacity computers and further understood that he had lost the punch card world to IBM with their 80 card system versus the Rem Rand 90 card system, and he felt that by acquiring EMCC, he would be able to leapfrog IBM and become dominant in the next era of business and scientific computing with electronic computers.

Figure 5: The UNIVAC ONE Worlds First Commercial Computer

A Difficult Transition

As it turned out most of the Remington officers and staff did not really understand computers since most had grown up with punch card equipment and could not crossover to the new technology. This posed major problems for Rand because he needed technology and marketing expertise to develop the company division and did not trust the two college professors and scientists, Eckert and Mauchly to reach the growth goals he had in mind. At the time, also, Rand had decided to hire several top retired military personnel, as he had earlier, by convincing General Douglas McArthur to join as Chairman of the Company, with the hope that they could open doors within the Government to allow him to promote the sales of computer and punch card equipment to various agencies in the U.S. government, primarily the Defense Department. One of the significant names hired was General Leslie Groves, a retired General who had been in charge of the Manhattan Atomic Bomb Project. Groves was immediately put in charge of EMCC, where one of his first acts was to demote Mauchly to the sales department because he flunked a security clearance test. The word was that he had joined and attended some Communist Party meetings in the 1930's). General Groves used to come to our Washington Office occasionally, and when he did, I was given the job of being his personal chauffeur, driving him to the Pentagon or one of the local Government buildings to give a speech or for various official meetings with top Government Brass. I got to know him well and although he was curt and short worded with me, he was interested in hearing me talk and listened to some of the dreams I dreamt. He must have been somewhat impressed because on one four hour drive to Fort Lee in Virginia, he mentioned to me that he had requested Jack Veale, my boss, to promote me to the Army Division of UNIVAC as a marketing representative. To this day, it is hard for me to grasp how as a young kid, I got to know and meet one of the key figures in American history that played such a huge role in winning World War 2 with the dropping of the two bombs on Japan that ended the war.

Everybody's Jumping In!

When I joined the Company in 1955, the industry was just beginning to explode. IBM had quickly recovered their dominance in the industry by designing and building a new line of computers that emphasized random access memory drums for storage versus old fashion magnetic tape, while UNIVAC which had leaped ahead with their earlier systems fell quickly behind. Groves had elected to move Eckert and Mauchly to the "cellar" and created a marketing strategy of selling fewer computers to improve quality, ordering production and quotas to be sliced by 50% annually. Morale quickly dropped in the marketing and management ranks and there was an immediate brain drain when top personnel left for other Companies, or to form their own. I watched with my own eyes the field of competition grow with Companies like NCR and Burroughs that had been in the bookkeeping world becoming computer startups, GE, RCA with a new series of solid state business computers called the 501, The Bendix Corp with their scientific "G" 20 series, Minneapolis Honeywell with a business 800 line, Scientific Data (SDS) with a powerful new scientific computer. With UNIVAC faltering, many talented people, like Bill Norris who left UNIVAC St. Paul, the major design and manufacturing center for UNIVAC, to found Control Data Systems (CDC) and offering everyone in UNIVAC, including me, a chance to buy stock at 75 cents a share. People around me in the office, friends like George Dancu and Dick Gasparie were leaving to join the new RCA Computer Company around the corner from us at 16th and K in DC. Anne Lamb, a brilliant programmer, whose father helped design the Burroughs' (Electrodata) first computer leaving and moving across the street to the Burroughs' office at 1631 L Street joining her father and his new team. Many of us, including myself, with less than a year's experience, were being offered jobs at higher salaries. I was tempted to take an offer with the RCA group, but at the last minute backed out and stayed. I thought with the talent drain, my chances of moving up in UNIVAC just got elevated, assuming they stayed in business. I mentioned several people above that had left, and one in particular, my friend Dick Gasparie, happened to end up being a significant person in helping me build my later successful career; and I'll spend time later talking a little about Dick. Groves with all his talents managed to move the Company's position in computers to an "also ran" status. By the 1960s, UNIVAC was one of 8 major American Companies in the computer industry then referred to as "Snow White and the Seven Dwarfs", IBM being the largest cast as Snow White and the other seven as being dwarfs! In the 70s', GE sold their computer business to Honeywell and RCA sold their computer division to UNIVAC, and the remaining dwarfs then became known as the "BUNCH" (Burroughs, UNIVAC, NCR, Control Data and Honeywell). A little bit of significant trivia happened that year, 1955 as well. Steve Jobs was born in February,

1955, and Albert Einstein died in April, 1955. One genius replaced by another!

Figure 6: Joint Chief of Staff General Maxwell Taylor
Remington Rand Trade Show

Early Computer History

It might be worthwhile to stop for a second and take the reader back into a little sequential history of the birth of computers. In a simple way, it's easier to break up the evolution and development of the industry into three phases starting with gigantic sized Vacuum tube computer systems, followed by smaller transistor sized systems, then ultimately the third phase that exists today, the micro systems, i.e. desktops, laptops and hand held devices such as the I-Pad.

When someone thinks about where computers were first created, the first thought that crosses your mind is California and other parts of the West Coast because that's where Silicon Valley is; Steve Jobs and Bill Gates lived there. Larry Ellison, Hewlett

Packard, Cisco, and others, mostly started and are located on the West Coast. Actually, the computer industry was born in the East, concentrated in Philadelphia, Norwalk, Boston, Upstate New York and Washington, D.C. Early companies like IBM were located in Poughkeepsie and Armonk, N.Y.; UNIVAC and EMMC in Philadelphia, RCA in Morristown, N.J., and others in the East building the earliest vintage vacuum tube and transistor computers in the late 1940s' and 50s'. Burroughs was an exception, although an east coast company, they acquired their entry into the industry by acquiring Electrodata Corporation located in southern California. When the move toward minicomputers in the 60s' started, the major firms that dominated and started first were, again, east coast companies, such as Digital Equipment Corp.(DEC) in Maynard, Mass. founded and run by Ken Olson. Ken started his career as a design engineer at MIT Lincoln Labs for a number of years before leaving to create DEC. By the end of the 90s's DEC was gone! First sold to COMPAC Computers and then finally to the Hewlett Packard Corp in 2002, and then Interdata Corp. In Oceanport, N.J., Jim Bruno sold to Perkin Elmer Corporation and then ultimately to the Concurrent Computer Corporation. Data General Corp. (DG), a Massachusetts company founded by Ed DeCastro, who worked originally as a key designer for DEC, left and started his own Company near DEC in Massachusetts. Data General lasted through the 90s'. When it was acquired by EMC in 1999, EMC's only interest was in the CLARION storage products of Data General and quickly spun off the computer piece to others. Sadly, the last vestige of the Company disappeared in October of 2009 when the old DG domain site http://www.dg.com was sold to the Dollar General discount store. By the end of the 90s' the mini computer industry had faded into ancient history! It is amazing that so much exceptional talent resided in the East and that they missed the micro-computer era so badly. Maybe it can be best explained by a comment made by Ken Olson in 1977 when he said "There is no reason for an individual to have a computer in his own home".

Looking back it was remarkable how quickly innovations occurred in the industry and how competition blossomed and spread across the country. As I have said before, I was fortunate to be a part of a new industry revolution that would be compared equally with the Industrial Revolution, the auto industry, and yes, Mr. Edison's light bulb and the Bell telephone! I was with Remington Rand UNIVAC and subsequently Sperry UNIVAC for twelve years and the number of new computer systems that UNIVAC developed in those few years was awesome. Just to give the reader a feel, I have listed here a few of UNIVAC "firsts" during those early years.

- ENIAC -First all-electronic digital computer
- UNIVAC-1-First commercial computer
- UNIVAC-1103A-First computer to use core memory
- UNIVAC Solid State-First all solid state commercial computer
- UNIVAC 1107- First thin film memory computer
- LARC- The World's most powerful scientific computer
- UNIVAC-490-First Real-Time computer
- UNIVAC File Computer-First Random Access computer

Early on, when first arriving at the Company, there were probably fewer than 200 people in the entire computer division, just a small part of the Rand conglomerate. The Razor division, for example, dwarfed the technology group. It felt like, and was, a startup, where you practically knew everyone in the Company by first name. I never called General Groves "Leslie", but other stars in the Company were known by first name. Dr. Grace Hopper, for example, always went by "Grace" and she even knew my first name! Dr. Hopper was a world renowned mathematician, who joined the Eckert Mauchly Corporation in 1949, and stayed for years with the Company when bought by RemRand. It was here that in 1953 she invented the first compiler called FLOWMATIC that translated English language instructions into the language of the target computer. She later said that the reason she built the compiler was because she was "lazy" and wanted to help the programmer return to being a mathematician! I was fortunate to have been a student in one of the FLOWMATIC classes that she later taught. Grace ultimately moved on from UNIVAC and her career was split between the commercial and academic world, as well as the military. For a while she was one of the original "Waves" and then later joined the U.S. Navy Reserve, but was called to active duty in 1967 to take charge of the Navy's standardization program, where COBOL, a business oriented compiler she had originally developed at UNIVAC, was being implemented. She was promoted to the rank of Commodore in a White House ceremony and two years later was raised rank wise to become the first female Rear Admiral in the U.S. Navy. One of the stories that Grace loved to tell her classes was how the word "debug" came to be. One evening the UNIVAC 1 she was working on had a power failure and when the technicians were locating the failure problem they found a large moth in the circuitry that shorted the system. There was a "bug in the system", so the circuitry had to be "debugged". So she coined the phrase "DEBUG" which we all still use today. Dr. Hopper died on New Year's Day in 1992 and was buried with full Military honors in Arlington National Cemetery.

There were many other "pioneers" that I worked with over the years at UNIVAC, that were not household names like Dr. Hopper, or Leslie Groves, or Drs. Eckart and Mauchly; but fellow workers that tasted the same early excitement I did as part of one of the major corporations that started our industry, and also folks who did pretty well on their own. People like Jack Veale, my first boss, who went on to run Optical Scanning Corporation, one of the first optical scanning companies in America, who later sat on one of my corporate boards, Richard Daly and Bud King, founders of Aries Corporation, one of the first software companies in the U. S. for whom I briefly worked, Dick Mayhew, a great guy who worked for our Air Force group, and later joined Computer Sciences Corporation (CSC) running one of their largest Divisions in Huntsville, Alabama when Werner Von Braun was at the Huntsville Missile Development Command. Dick later left CSC and founded a publicly held software company called General Computer Services, in which I became one of the first stockholders. Vince Grillo, also part of the Air Force group, and perhaps the most charismatic "marketeer" I have met during my long career, who left UNIVAC and joined CSC, and, in my opinion, was the star that catapulted CSC into becoming the largest software company of the 70s' and 80s'. Fletcher Jones and Bill Hoover owed much to this man for his overall contribution and the many major contracts he personally won for this fledging company. There are a number of legend stories about Vince, both at UNIVAC and CSC. The one I most remember at UNIVAC was; he had to make a flight to New York in the mid 50's and rushed to catch a shuttle on Eastern Airlines at National

Airport, now Reagan, and forgot his wallet. Back then there were no security checks, so he went directly to the gate; convinced the attendant to let him board the plane, with no way to pay for his ticket; took off, landed in New York, and then miracles of all miracles, with not a dime in his pocket, persuaded a NYC cab driver to take him to the City and borrowed money from his customer to pay the cabbie, to get back home. Now that's salesmanship! But an even better story about Vince, I'll tell when we get to the CSC years. When I left UNIVAC, I followed him to CSC, worked directly for him and was given the responsibility for building the "first" formal worldwide commercial and government marketing organization for the Company. More on that a little later as well!

Moving West!

It was not until the late sixties, with the startup of the microcomputer world with the IBM micro announcement (remember the old Charlie Chaplin pc ads) that the West Coast emerged as the leading edge designer and developer of hardware and software technology. Firms like DEC and Data General missed the "pc boat" and one by one died off with most of their personnel moving to the microcomputer industry. Many of the early key designers and marketing talents left to go west to new startups, and "old timers" such as Hewitt Packard, Mini-Computer alumni like, Ray Ozzie, a key software developer at Data General ended up replacing Bill Gates as the Chief software architect for Microsoft, (until he left at the end of 2010), along with Craig Mundie, now the Chief Microsoft Technologist. Jonathan Sachs, also one of the major software designers at DG moving west where he then co-founded Lotus Development Corporation, authoring the popular Lotus 1-2-3. To Apple Computer from DG France and "Be" Computer went Jean Louise Gassee, who for a while was being considered as a replacement for Steve Jobs. Fat chance! Steve didn't believe the words" replacement or successor "should be in the Apple dictionary! Len Basak, a key figure within the DEC Research Lab in Massachusetts rode west to co-find CSCO Systems with his wife, at the time, Sandy Lerner. A side note on Sandy, after leaving CSCO, she decided to go the opposite way and moved east, ending up in

Upperville, Virginia, just a mile or two from where I reside in Middleburg. She decided to leave the computer industry, with her billions, and bought a farm; started raising organic animals and crops; purchased a grocery store selling her produce and meats, and built an English Pub in the center of Upperville, where she promotes her organic food! It probably ended up being a larger challenge than starting CSCO, since the Upperville town loudly protested and fought the Pub for a lengthy period of time. People are funny in this little elegant and "sleepy" town where Paul Mellon resided, the motto "no development, no commercial exploitation and stay green" is a running theme thru the countryside. But what the hell with a few billion to toss around anything is possible! Today, the Pub is thriving and if Paul Mellon were still alive, I'm sure he would be dropping by occasionally to down a "pint or two".

"The New Recruit"

When I first arrived at the Company, I was introduced to a likable Irishman by the name of Leo O'Keefe, a brilliant computer hardware and software genius and placed under his "wing" initially to indoctrinate me into the "ways" of the company. Leo had previously worked for the Government at the National Security Agency (NSA), one of the earliest users of computers, and he had been hired by UNIVAC to teach employees and customers how to program and use the large scale UNIVAC 1 and UNIVAC 1103 computers. He also was an extremely competent writer, and developed a number of manuals for these early systems, as well as from time to time, he was used to write difficult and complex sales proposals for a variety of government and commercial prospects. Leo was a graduate of Harvard; spoke fluent Japanese, and was a college classmate of Henry Kissinger. He absolutely was one of the most brilliant and entertaining people I ever had the pleasure of working with, and without question an individual of many talents. But interesting enough, his real love was not the computer industry, but entertainment. Leo loved to play the piano and had a fascinating singing voice that sounded identical to that of Mel Torme, one of the famous crooners from the same years as Frank Sinatra and Perry Como. Leo loved his "whiskey and milk", the milk was for his ulcer, and he indulged himself after work and on weekends by playing the piano and singing Mel Torme songs in his favorite cocktail lounges around the City, where he was popular and worked for free drinks. One of his favorite hangouts was the "Fireplace Lounge" on 18th Street in Washington that was a popular local cocktail lounge. Just a tiny cabaret where a bunch of us would hang out to listen to our friend Leo sing his favorite "Mel" songs and listen to, believe it or not, Roberta Flack, who was a student at Howard University majoring in music and singing part time at the Fireplace. The Fireplace is where she got her start, and from time to time when Leo was there, he would accompany her, playing the piano, while we would hang around the piano listening to her sing. As she became more popular, she moved to a larger setting in a club on Capitol Hill, where Leo also played, and from there she moved on to become one of America's beloved singers. Leo went back to Computers, and the two of us worked together until the day he retired!

Before I was assigned to what they called the "Army Pool" Division of the Company, a position I received thru General Groves recommendation, I spent most of my first year being sent to a variety of computer hardware and software schools as a young trainee, So schooling took place in a variety of places, including the Washington office at 1630 L Street, for basic training; the "Brassiere "factory in Philadelphia, at 18th and Allegheny, (at one time a women's lingerie factory) for software training; and to the Remington Rand UNIVAC formal training school at Fischer's Island, N.Y. All of the training in D.C. and in Philadelphia was extremely challenging and interesting, because most of it had to do with all of the latest computer technology and systems. But when I arrived at Fischer's Island along with 25 other new recruits, the school training was based on punch card and old fashion calculator computers, like the UNIVAC 60 and 120 systems. I lost complete interest in these classes and almost my job, ending up close to the bottom of the class at graduation time. I still vividly remember the parting interview I had with the senior instructor, when he looked at my grade scores; stared at me and said, "Jack, what did you do before you came to UNIVAC"? When I answered that "I worked for the American Can Company", he replied "you'll never make it in the computer industry; you might think about going back to tin cans and milk cartons!" The really significant thing I remember about that last day, as I write, is that Fischer's Island was truly an island across from the mainland of Connecticut and you had to take a boat to get to New London, Connecticut, which a number of classmates would do, when we would get weekends off. Well, the Connecticut School for women was almost directly across from the school, and on weekends we would meet some of the women at the local bar near the school, and we were always teasingly invited to come visit them at the dorm after we left them at the bar. One night, just two weeks from our graduation, the invitation was just too inviting; so one of my friends in the class from the Baltimore office, Darwyn Kelly along with myself, decided to do just that; and we did! Except to get onto the campus we had to scale a huge wall since the main gates were locked. It was a struggle but we made it over the wall and it turned out to be a wonderful night, except for one thing. I caught the worst case of poison ivy I've ever experienced scaling that wall and on the day of graduation, at that moment, didn't know what was worse; being told I was a "loser "and would never make it in the computer business or the agony of a body covered with poison ivy!

Figure 7: Remington Rand/Univac Training School
Fishers Island Connecticut

The office in Washington where I worked was a mix of people operating in the commercial computer world along with a government division. Additionally, there was a training department, as well as offices for our lobbyist group and visiting dignitaries like General Groves, or even Jimmy Rand, himself, on one of his infrequent visits to D. C. The Government Division was broken down into what were called “Programs" by military or non-defense activity descriptions. For example, there were separate programs for the Navy, Air Force, and the Army, as well as a program for each non-defense activity, such as the FAA, NASA, Department of Agriculture, etc. I was assigned to the Army group, because of my connection with my old "pal" General Groves and reported to work under the watchful eye of Dale Holmburg, my new boss, early in my first year at the company. Initially, I began as a system support analyst helping clients program and use their computers and accompanying marketing people on sales trips to different Army bases around the country; basically just "learning the ropes". In between, I attended marketing meetings with managers and others at UNIVAC Headquarters in St. Paul, Minnesota, or even a little later meeting Mr. Rand, himself, at a corporate party held at his home in Darien, Connecticut. I specifically remember the party at his home was to introduce Dawes Bibbe to the marketing organization. He had just been recruited from IBM to replace Groves as the President of the UNIVAC Division, and the excitement that had created was high. Bibbe was a key senior executive at IBM, and it was a coup to land such a recognized name in the industry, when UNIVAC was reeling and falling quickly into "an also ran" category against the rising tide of the IBM bemoth. Sadly, in the end, He was never able to turn that tide and make UNIVAC once again number one.

That first year moved rapidly and once assigned to the Army group I was placed on a support team implementing a UNIVAC File Computer (UFC 0) at the Military Transportation Agency located at Gravely Point, Virginia, which was at the site now known as Crystal City, a mix of Government and Commercial offices and residential buildings. The UFC 0 Computer was a newly developed general purpose system, a forerunner of the UFC 1 to arrive later in 1956, which featured an easier to use three address programming system and massive random memory access drums for large amounts of storage, which the military Agency needed. The project was a disaster. Much of the computer system still had early design bugs that were being corrected slowly at the customer site and it was a slow bitter process, before the Army finally cancelled the project and replaced the UNIVAC with a newly announced IBM RAMAC 305, their version of a mass storage drum storage computer. The industry was rapidly moving from old UNIVAC 1 technology, which mainly used magnetic tapes to input and output data, and converters to transfer punch cards to magnetic tape, and vice versa, to technology that used random access memory drums to move and store data internally. IBM's answer was the RAMAC 305 and it took the world by storm and became the preferred medium scale system for many years. It was the system that moved IBM into the number one computer position, after moving slowly from punch card technology to computers. The senior Tom Watson still believed in punch cards - as he kept saying "you can see and feel data in punched holes, but you can't see it on tape or an internal drum device". It took him awhile to change direction. As Steve Jobs would have probably said "Thank God for younger generations of kids that come along with new ideas to do what can't be done".

At any rate I was moved from Gravelly Point back to the Washington office and was assigned to the Army marketing group, where I wrote technical proposals for potential new business opportunities. At that time, except for a few large scale installations, the U. S Army, unlike other Defense Agencies, was a user of mostly punched card equipment, including the smaller UNIVAC 120 computer. Further, most of the Army used IBM equipment, so it was an uphill fight to convince them to switch from using the IBM 80 column to our 90 column card, since most of their application work had been standardized on IBM equipment over many years. So we would look for, and bid, on selective opportunities that were not previously mechanized. One area the Army had to decide to automate was their various training schools, which we were successful in winning, by including newly developed optical scanning equipment for test scoring; placing equipment at various training centers around the country; the Aviation School at Ft. Rucker, Alabama, the Signal School at Ft. Monmouth, N.J., the Infantry School at Ft. Benning, Ga., the Southeastern Signal School at Ft. Gordon, Ga. and Ft. Benjamin Harrison Training Command in Indianapolis.

As part of these contract wins, the Company had to supply program management support to each of these areas, and since I was one of the "single" people in my group I was assigned initially to the

Ft. Gordon installation as the support manager for the better part of that year. Imagine that! Escape the draft and end up at a military base for the better part of a year! Only difference it seemed was I wouldn't have to wear a uniform! Ft. Gordon, Georgia was located in the beautiful little town of Augusta, home of the U.S. Open and a town I learned to love, except the person I loved was still in D.C.

"Relocation"

So for the last several weeks before I left for Georgia, Pat and I spent quite a bit of time together, but movement in our lives was shifting and it was obvious that maybe what was meant to be was not what was meant to be! So sadly we parted. Her parents decided to move to St. Petersburg, Florida and Pat moved into an apartment just over the District line in Maryland, and I packed up my Ford convertible and headed off to Augusta. There were also an eventful series of changes at RemRand, as well. Mr. Rand had decided in the past year, 1955 to merge the Company into Sperry Corporation, and General Douglas McArthur, at 75, was named Chairman of the Company; Rand became Vice-Chairman and Harry Vickers became the first President of the Sperry Rand Corporation. Companies in the 50s' were merging and entering the new computer industry with the combined hope that new products and more powerful sizes of acquisitions would perhaps stem the growth and monopoly of IBM or eventually replace them as Number One. Amazingly, Remington, started in 1873 ;and the Company that built the World's first commercial typewriter; merged with Rand Co. in 1927; a maker of "Rand-Kardex" record keeping systems with revenues of $10,000,000 ; next acquiring Eckart Mauchly, and now a part of Sperry Rand Corporation, a new Company with combined revenues of $484 million. It was almost a merger of "equals" as Sperry

Figure 8: Taking a Break with Pat and Wally Mitchell

exchanged 3 1/2 shares for 1 in the new Company and Rand 2 shares for one. By now, the hope was that the combined Company should become a formidable competitor to the "Giant" in the business, "Snow White"! Sperry was a Company that by the 50s', had acquired an outstanding reputation for their work in hydraulics and electronic instrumentation and early rocket and missile development work. A long legacy in electronic research and development, stretching all the way back to the 1930s working with Lt. Colonel James Doolittle creating the first horizontal horizon and aircraft directional gyro. Who could lose with names and reputations behind you like Rand, Doolittle, Groves, McArthur, and Eckert-Mauchly? Well, have you ever heard of Governor Dewey?

I only digressed a moment from talking about my breakup with Pat and my migration to Augusta to describe a few of the major changes happening in my life and the Company and worrying about how they might possibly affect me. The old saying "out of sight-out of mind" kept cropping up in my head. Heading to Augusta would anyone ever remember me, or when I returned would I have lost my place in the promotion hierarchy! I read a recent interesting article in the Wall Street Journal describing unemployment in the U.S. and how many technology employees, who work from home by telecommuting, were all of a sudden voluntarily changing their work environment and heading to the office every morning! Insecurity does rear its head, and anyway, it was kind of a thought going through my head as I was driving south to Georgia. Maybe as a backup plan in Georgia, I might have to take up Golf as a profession; become a golf pro and one day teach "Tiger" how to make a living!

Moving to Georgia was a monumental change in my life. For a kid who had only spent brief vacations or business trips away from home, this was serious stuff! On my own; in a new town; didn't know a soul; solely responsible for support managing a large data processing installation at Ft. Gordon. Wow! As soon as I arrived, I initially moved into one of the Officer's BOQs' at Georgia and shortly thereafter took up residence at the old Bon Air Hotel in Augusta, a beautiful spot that many vacationers stopped at when coming to see and enjoy the U.S. Open. For the next 9 months that was to be "home". Augusta was a lovely little town back in the 50's and was known by most locals as the home of "old money". Many of the wealthy had settled in Augusta generations ago and built large ante -bellum mansions that dotted the local town from end to end, But aside this obvious wealth were the poor, mostly Negro population, living in shanty housing, coupled with a section of the town dedicated to the military population stationed at Fort Gordon, home of the Army Southeastern Signal Command. It was at this base, where soldiers were taught how to use all of the latest state of the art equipment including telephony, radar, surveillance and satellite communications.

Founded in 1736, Augusta was awash in early history of the U.S. There were many historical sites to see and visit in my spare time; places like the old Cotton Exchange, which used to be the second largest in the world when it was operational and the old confederate civil war arsenal, located on the campus of Augusta State University. Gorgeous trees blanketed the streets and walkways throughout the town, and you could get a beautiful view of the Savannah River, adjacent to the town, by walking the Augusta Riverwalk. Many famous people have lived here over the years, including George Walton, the youngest signer of the Declaration of Independence, who spent a good part of his childhood there. Also, Woodrow Wilson, lived here as a child, when his father was Pastor of the First Presbyterian Church. And of course, I lived here!

Augusta, although beautiful, was a sleepy little town, with not a lot to do, for a young 20 year old, unless you played golf. So mostly my time there, other than work, was an occasional dinner with one or more of the employees from the computer group, fishing for "cats" on the Savannah river, strolling the town, visiting historical sites, seeing a movie or two, or even a play or show at the old Imperial Theatre located on downtown Broad Street. The Imperial Theatre had been built back in the early 1900s', right after the huge fire that had raged and destroyed a large part of what was then Augusta. Once built, it attracted a lot of well-known burlesque and movie stars that performed there. People like Charlie Chaplin, who sold Liberty Bonds in 1918 by performing onstage. Also, a well-known popular silent movie star at the time, Leo Carrillo, who played "Poncho", the Cisco Kid's partner, was known to have played the Imperial Stage.

The months passed, and I would occasionally get a letter from Pat, telling me "tales from the hallways" of her elementary school;

how much she enjoyed teaching, and although now dating others, still writing she loved me. Also, she had moved, deciding it would be safer and less expensive to move in with one of her old classmates, Janet Trebach, who was also a high school teacher in D.C. She came to visit me in Augusta, early that summer, and things started to "click" again! So by now, we were convinced that we really belonged together; and on a trip back to D.C., late in the summer of 1956, we decided to get married in 1957, but not before she caused a minor uproar in the Catholic Church! That story, I'll tell in a second or two. Shortly after she left, I was told to wrap up my work at Fort Gordon and report back to D.C. In September, since I would be needed to further market and promote educational systems to other Army Training Commands. I forgot to mention, that while she was visiting that weekend in Augusta, we happened to pass a car dealership that was selling a new BMW import car called an Isetta, a small two seat car, really almost a motorcycle, on four tiny wheels with one door that opened at the front of the car. She immediately fell in love with the little yellow "bumble bee", and I ended up trading my beloved, baby blue, Ford convertible for this tiny Isetta bug! Sid Barteau, my old boss at CANCO, would have thought I was crazy. September soon came, and I started north in my new yellow bug, with the two and a half cylinder engine, hoping I would make it back by Christmas that year. I always kiddingly tell the story that it only took 5 gallons of gas to get back home, but a year to get there! Later that fall, Queen Elizabeth came to D.C. to visit the White House, and during her visit she attended a football game between my old Alma Mater, the University of Maryland and the Naval Academy, where she arrived in a parade of secret service and embassy cars, and bringing up the rear end of that parade were the two of us in our new "Isetta"! Barteau would have loved that!

"Surprise -I'm back"!

Once back home, and still single, I moved back in with my parents, whom I mentioned earlier, had moved to Maryland from the City. My brother Phillip, discharged from the Navy was also living with my parents, and as he tells the story he was lucky to reconnect with the family. When he was discharged, he drove back to the old house in D. C., and with his duffel bag in one hand, and the house key in his other, he attempted to open the door. To his surprise, the key wouldn't work, and after ringing the bell, a woman he didn't recognize, opened the door, and announced she was the new owner! My parents had forgot to let him know they had moved. There is an old saying "that the youngest child is sometimes lost and forgotten in the crowd"! He didn't stay long though, a couple of years later, he married a girl from the City, Eileen O'Connell, and went on to a successful career running the D.C. City Sanitation Division, reporting directly to Mayor Barry, who by now was a national household name to most Americans, for his wild and reckless personal life that he has led for decades. Drugs, prostitution, jail time, failure to pay taxes; you name it and Barry did it! My brother, successful in his profession, in spite of being in Barry's circle, has had a happy and gratifying family life. Married to Aileen for close to 50 years, they have raised a family of six kids, and at this reading 13 grandchildren.

Things had changed dramatically when I returned to the D.C. Office. Remington UNIVAC had merged with the Sperry Corporation and the name changed to Sperry UNIVAC or to most, the name became simply "UNIVAC".

Figure 9: With Lt. General Campbell (left) and Carl Knorr, Remington Rand Univac

Jack Veale, the vice president of our division had left to take over as President of the Optical Scanning Corporation, in Newtown, Pa. and was replaced by a transplant from Corporate Headquarters, Carl Knorr. Many newcomers were now working at the Company, and this new, first generation of computer employees, was now poised to change the way that business and individuals handled their business, scientific and personal work. Truly an exciting time! In addition to the Washington UNIVAC office expansion, a large number of employees were added to the branch offices around the country, which were used to assist us in the support of Government computer contracts awarded at locations around the country, principally in the South, where there were a large number of operational military bases. Word soon spread thru the Army Training Centers of our success at Fort Gordon and we were soon asked to bid on a number of these to provide them with a competitive bid, against mainly our "friendly" competitor, IBM. Since, I had been mostly involved in the contract at Ft. Gordon, I became the lead marketer assigned to pursue these other opportunities, and for the balance of 1956 into 1957, most of my time was spent traveling in the South, focusing on potential business at Ft. Benning in Columbus, Ga., not far from Ft. Gordon, and Ft. Rucker, home of the Aviation School, in Dothan, Alabama.

Getting back home from the Ft Gordon assignment, also gave me a chance to reunite with Pat and old friends, as well as my family; so we had plenty of time to start serious discussions about the "Big Day" scheduled for next summer. My sister's husband had been earlier discharged from the Coast Guard and was studying at Georgetown University under the GI bill, so we got to spend time with them, as well as old friends from college - friends like Ray and Anne Lunceford, the Kendalls, the Lamons, Olsson's, Janice

Treback and her soon to be husband, Jim, and others. Lunceford was now working for the computer division of Royal McBee Corporation, a new entry in the ever growing number of Computer Companies "popping up" everywhere. By the end of 1956, there were close to 25 companies that had, or were building computer systems, including peripheral devices such as magnetic tape, drums, core memory, and card to drum, or card to magnetic tape units; manufactured by Companies, such as Mohawk Data Systems, a popular manufacturer of these faster hardware converters. On a different business front, my old friend, John Olsson was discharged from the Army and on his way to building one of the largest chains of book and record stores in the Washington area! It was also a "hit" year for Broadway musicals with the opening of "My Fair Lady" starring Rex Harrison and Julie Andrews;" Bells are Ringing" with Judy Holliday, Sydney Chaplin, and Jean Stapleton, later of "Archie Bunker" fame, and lastly," Li'l Abner", with the fabulous Edie Adams playing Daisy Mae with lyrics by Johnny Mercer, one of my favorite musical writers. Who from that era does not remember the song he wrote earlier, "Old Buttermilk Sky" that he also sang and that rested at the top of the charts for weeks on end. It was everyone's favorite! I did get to New York before Christmas to see Li'l Abner with John Olsson and another friend, Jack Carr. S ince we were all working, we escaped staying at the "Y" and found a room at the old Roosevelt Hotel, at 45th and Madison, with our own bath, certainly a move up in the world!

"New Challenges"

Fort Benning became the next target for us to tackle, but this time I had the additional assistance of two outstanding associates from our Atlanta office, who were assigned to help in convincing Fort Benning to switch from their IBM equipment to ours. E.W. McCain and John Imlay were two bright and "Gung Ho" young guys, who like me, had joined the Company in recent times and were anxious to prove their mettle! So, the three of us blanketed the key military evaluators' and ultimately won the contract. We spent a ton of time at Fort Benning; flying D.C. 3s' from D.C. and then from Atlanta to Columbus, on and off for several months, getting to know each other quite well. A little on Fort Benning; It was the major infantry base for the U.S. army, located in Columbus, Georgia, where soldiers trained and attended "boot camp", before being sent to their next duty assignment. Servicing the military population were also a large number of strip malls providing the usual fast food diners, bars, cut rate gas stations, pawn shops and entertainment usually found near most military installations in the States. Speaking about these" strip" malls, across the main bridge connecting Columbus was Phoenix City, Alabama a wide open town that serviced the military in another way. Phoenix was a small Las Vegas, with "strip shows", prostitution, gambling, seedy bars and virtually anything else that a "needy" soldier may need and was sure to find! Virtually lawless, it was run by a crime syndicate led by Jimmie Matthews and Hoyt Shephard that controlled the city. Margaret Barnes, a well-known author, wrote a book on Phoenix City, after the Attorney General, Al Peterson was gunned down outside of his office, called the "Tragedy and Triumph of Phoenix City". For years, the town was known as "Sin City, I believe that a film was made later on about the City, based on the Barnes book. I did win this significant contract, and several others, thanks to the help from both E.W. and John, and we went on to share many additional "wins" together, although E.W. (Mac) and I did have our problems and strategy differences; often separated from arguments by John Imlay, the "junior" guy in our midst. John wasn't a "junior" for too many years, however, as he turned out to be a tremendous success in our industry and honored in 1997 by being voted into the Technology Hall of Fame for Georgia. During his career he started several companies, including Management Science America (MSA) one of the top ten largest software companies in the world with over 60 offices worldwide. He sold this company in the 90's for nearly $400 million to Dun & Bradstreet. Under his guidance, the company birthed over 300 CEOs' and some 100 companies in the process. Always a football fan, John became a personal friend of Fran Tarkington, a one-time quarterback for the Atlanta Falcons, and later became a business partner with Tarkington, as well as purchasing an interest in the Falcons team, several years ago. In recent years John has backed a total of 33 startup technology companies in his lifetime and is considered by many in Atlanta as the "Godfather" of Technology in Georgia startups. "Mac" McCain came from a wealthy Georgia family, whose lineage included the Howard Coffin family that founded Sea Island, back in the early 1900s'; Mac attended the Citadel, and his military education helped immensely when we dealt with the military. He left UNIVAC in 1963 to join Sam Wyly's company, University Computing Corporation (UCC), a time sharing computer service bureau business formed to compete with the IBM SBC, (Service Bureau Company) selling computer services to users that required off site additional speed and computer capability. It was less expensive for a customer to "buy computer time", rather than install their own system. UCC used large scale UNIVAC 1107 and 1108 large scale main frame computers in their centers, and McCain, coming from UNIVAC, was familiar with their capabilities. He was put In charge of the marketing and sales operations at UCC, thru 1972, when Wyly's split the Company into four; one of which became Datran, a Company that built nationwide microwave towers in 27 cities to compete against AT&T. I lost track of McCain during those years, but Wyly, who I never met, still is a significant entrepreneur in the computer and software world, as well as in other businesses that he owns. In addition to founding Wyle Corp., Sam acquired Gulf Insurance, Co., Computer Technology Corp, the Bonanza Steak House Chain, which he sold in 1988 after building it to 600 locations, bought controlling interest in "Michaels", an arts and crafts chain, and co-founded Sterling Software, which was later sold to Computer Associates in 2000 for close to $4 billion. One of his latest investments is in the social networking company, Zaadz.com, in which he is the single largest investor. Not bad for a kid who grew up in East Carroll, Louisiana, one of the poorest towns in the U.S.! On a minor note, when I joined CSC in 1966, I was one of nine employees, including the President, Bill Hoover that developed the business plan for a service bureau company that would compete with both IBM (SBC) and UCC. Computer Sciences exited that business a few years later to focus on their software world, when they found they were not competitive with Wyly and SBC. Later, my European Company, CDSIL, started a time share company in Geneva and Paris, and our Paris Installation was located at the site of the earlier UCC planned location.

Before I leave the Fort Benning story, I should mention that it was there I met a young enlisted man by the name of Walt Scott, who was one of the senior Army programmer's assigned to the computer project. Walt was smart and eagerly wrapped himself up in the project, and contributed greatly to the success of the installation. When he was discharged from the Army, He moved back to D. C. and I hired him to work for UNIVAC. Later on, he joined me at CDSI, as one of the first six employees at the company, and also one of the key people who was a significant player in the successful startup of the company. Walt was a unique individual, who "crisscrossed" between jobs in both the commercial, as well as high level Government world; so he knew both arenas, and was an effective and successful communicator and problem solver between the two. He understood "both sides" well! Further along in his career, he once again joined me at C3, Inc., in a Vice President role, where he continued to provide invaluable support to both the Company and the Government on various contracts we enjoyed.

"Wedding Plans"

Spring of 1957 arrived quickly, and by then Pat and I were beginning our wedding plans, or I should say, "She was doing the planning-I was doing the listening". I had given her a ring the previous Christmas, and from there the only "hitch" was the qualms about religion. As I mentioned earlier, if one were to marry in the Catholic Church, and were of another Faith, then you would have to sign a document agreeing to raise any children by that marriage as Catholic. Further, the non-Catholic party would have to take lessons in the Catholic faith, prior to marriage; and lastly, the marriage could not be consecrated at the altar, but would take place in the Sacristy, which was at the "back of the Church". If you knew Pat, you would know that she did not take to this quietly, or gracefully, but in the end "reluctantly"! So in late spring, she began "catechism" lessons in my local parish Church, St. Camillus, in Silver Spring, Md. with a new, young priest, Father O'Sullivan as the instructor. I'll never ever forget Father O'Sullivan! He had recently arrived from Ireland, where he had been ordained, spoke with a heavy Irish brogue; so heavy that I could hardly understand him when he spoke, and he had an extremely serious personality, and I do mean serious! Well, the class of five that she attended was not just for newlyweds" to be", but also for converts. People who truly wanted to convert to the faith, and were taking lessons because they were interested in learning more about the religion, with the ultimate plan to maybe join! So the class begins and for a couple of weeks everything seems to be going well, except for the moment I get a call from Father O'Sullivan, asking me to show up for a private meeting, with just him! When I arrived for the meeting, I could immediately tell he was distraught, and as he started to talk, I thought to myself "I should have brought an Irish translator with me"! "I can't understand the man"! Little by little, in a frustrating voice, I realized he was not very happy with "Pat's attitude" in class, and proceeded to tell me that she had become a disruptive voice in the classroom. Well, knowing her like I did, I could have told him that upfront and saved the man an ulcer. It turns out that she was asking a lot of direct questions on issues like "why were Popes allowed to marry or have mistresses in the mid-centuries; why was missing mass on Sunday, a mortal sin" and many other issues not worth mentioning here. After he finished, he turned to me, and in his Irish brogue said "Jack, I'm going to have to ask her to leave the group; she has the rest of the class that are here for conversion beginning to doubt". Thankfully, I guess she spent enough time in class to comply with the rules, and after signing papers relating to raising any children we might have as Catholics, she passed "muster". With all of that accomplished, we had cleared the decks for getting married that coming June.

Figure 10: Wedding 1st Dance

And June came! We were married that month, in the "back of the bus" Sacristy of St. Camillus, our Parish church, by none other than Father O'Sullivan! He had been so intimidated by Pat, and her arguments, that I was concerned that she might have influenced him to the point he would "jump ship" and suggest we all go around the corner to the local Episcopal Church and consummate the wedding there! It was a happy day for all; family including both sets of parents and my Grandparents; friends, and college classmates; Carl Peters, my best man and Jean Mitchell, Pat's maid of honor. Afterwards, that afternoon, and evening, we had a reception at the old Shoreham Hotel on Connecticut Avenue in the city; and the next morning we drove to St. Petersburg, Florida-----not in the Isetta, my back couldn't take it. ---but in her Ford sedan. We stayed at her parents' home in St Petersburg, for one night (the night of my asthma attack), and the next day drove to Key West where we caught a Cubana Airlines plane to Havana, Cuba ; yes, Cuba had their own airline at that time. Here we spent a honeymoon week, staying at the beautiful Hotel Nationale, touring the City, its night clubs, casinos, Hemingway's house, and the beach. It was a pre-Castro time in Cuba; the people seemed happy and busy and there was a spirit of gayety that seemed to hover over sun lit days and carried over to night when the City ignited in sparkling light. By the way, since the first night in St. Petersburg, I have never experienced another attack of asthma. Some say that marriage heals---others say----well, we better not go there! But it worked for me. I can see why Hemingway loved Cuba, and I can only hope that one day, our Country will once again allow citizens the right to visit. Our Cuban honeymoon reminds me of a story about several friends of mine, in the computer industry, that went to Cuba, on a supposed trade mission, which was allowed, back around the year 2000. Each of us worked for, or owned, separate competitive companies, but once a year, we forgot about computer competition and concentrated on "fun" skiing together out west. One friend, Tom Hewitt, a well-known and successful entrepreneur in Washington computer circles, decided to buy a good sized sail boat and sail it to Cuba from his home in Florida, under the guise of a computer trade mission to Cuba. All he needed was five volunteers to go with him, sailing experience preferable, but not really important! In his case, he had none, except for a couple of days being taught how to sail by the original owner. As expected, none of the other five, except for one, had stepped on the deck of a sail boat, much less sailed one. Away they went, and the post horror stories were downright scary; the storms, the winds, boat nearly capsizing, bodies being tossed everywhere; a "Perfect Storm" movie. Well, they made it, but on the way back, four of the six decided to catch a plane in Havana to

Toronto, and then back home. Tom always had a new side venture planned the following year we would ski, but nobody wanted to put his new suggestion on their bucket list of "50 places to see before you die", because they knew they would not see the next 49!

"Back to work and school"

On our return from Cuba, Pat's friend, Janet, whom she had been living with, suggested we continue to live with her until we got settled, which we did for the balance of the summer. Quite an adjustment for me! From "mothers" house to living with two women was, well what can you say? "It was certainly different", especially with one bath and two women! That fall we moved to an apartment in the Silver Spring area, and went about "playing house". That fall I returned back to night school at American University, working on a graduate degree in Political Science and continued working at UNIVAC during the day. American was the only local university at the time offering graduate computer classes, but they were being offered only under the Political Science program. Additionally, at the time, the School used part time instructors that were full time Government employees, with computer experience to teach, since at the time the industry was new to all, and there were few qualified full time teachers around. Certainly a different world back then!

Pat, shortly thereafter was pregnant with our first son, Glenn, but continued teaching most of that school year 1957-58, while I continued working with the Army Group, and school at night. Other than an occasional field trip, I was able to concentrate on marketing to the Central Computer Evaluation Command, in Virginia, where major Army computer selections were made, along with selective sites that were targeted accounts around the Washington area. The U. S. Army, by 1958, was locked into using IBM equipment, and it was extremely difficult to break into this mostly IBM "80 column" punch card market, since we made 90 column non-compatible equipment. Further compounding the difficulty was that the Army, except in isolated instances, was not interested in newer, larger computer systems, like other Government Agencies. So, we placed all our energies and focus on those few sophisticated sites that needed and wanted larger, more advanced systems. The world then belonged to IBM and it was an uphill battle to convince prospects that we had a better product. Fortunately, UNIVAC had announced a new system, the UNIVAC File Computer-1, which caught the eye of several Army Agencies, like the Chemical Corps and Transportation Commands. After major exhaustive competitive bidding against others, including IBM, we captured these Commands, placing UNIVAC File Computers at the Edgewood Chemical Arsenal in Maryland, and multiple systems at nationwide Army Transportation Centers. These sales and installations, including school, occupied my time for the next year well into 1958, the year my first son, Glenn, was born.

Once we found out that Pat was pregnant, we decided it was time to leave Janet's apartment and find our own; which we did, moving into a townhouse in Northwest Park, in Silver Spring in late fall of 1957, Meanwhile, my sister was also pregnant with her second child, "Laurie", while my brother and his wife were building their own family which ultimately included four children. It also would turn out that the Pat and I would spend approximately another twelve years living in and around the Silver Spring area, an area close to the University of Maryland. It would also be another nine years before I would start Computer Data Systems, my first computer company, at the age of 35. At that age, young Zuckenberg, of Facebook fame, would have started several successful "social" companies and long been retired! Meanwhile, it was back to work; with another pending "mouth to feed" and I focused on my job at UNIVAC, hoping to climb the promotional ladder. Since I had now been in the computer business for several years, I knew our product lines, as well as the major competition, quite well, and along with others began to successfully sell some of our larger systems into the Government marketplace. In those early years, I became quite friendly with several other young "marketeers" and support personnel at the Company; names like Lee Johnson, who handled sales to Non-Defense Agencies, such as the Dept. Of Agriculture and the General Services Administration; "Bud" King, Director of Software, "Dick Daly, General Manager of the Government Division of the Company, "Dick Gasparie, commercial marketing and Vince Grillo, Air Force marketing, Tom McCaffrey of our Richmond Office and Dale Holmburg, my immediate boss. These were all people that influenced and assisted my career in different ways, over the coming years; some of whom I briefly talked about earlier, and others that will come a little later.

Changes happened fast over the next several years; my son Glenn was born during the summer of 1958 and Pat did not return to teaching in the fall. She became a "stay at home" mom, which suited her perfectly. She loved children and was a "natural" when it came to understanding and raising children; and she reared our four children, lovingly and firmly, instilling her good values in each. We stayed in the townhouse for a year and then moved to a small bungalow home, close by in a tiny suburb of Silver Spring called Oakview. It was a perfect starter home that we purchased for $10,000 and assumed a 4 percent loan, after a small down payment. I vividly remember our monthly mortgage payment, including principal, interest, and taxes were $78.16! Remarkable, the little things that stick and remain in one's head! The Development had been built by Carl Freeman, a local prominent builder, who built homes with a contemporary look; and like the New York/New Jersey Levitt "cookie cutter" homes built in the 40's and50's, all of them looked alike. Oakview was a perfect spot! Neighboring families were young and friendly; it was close to work and nearby schools, and convenient to a number of our friends and families. Our next two children, Steve and Christopher were born, while we lived here, and Pat's brother, lived close by, and shortly remarried a young woman Debbie Ennis from Baltimore. Soon after, Jim and Debbie had their first child, a daughter, they named Suzanne. Over, a few more years, they completed their family with the additions of Paul, and Jo Ann, the youngest. The years began to go by rapidly, and with busy days and sometime nights at work, coupled with an expanding family, I dropped my night school classes, just a few credits and a thesis away from my masters. Maybe it wasn't all about being busy with family and work; I had been moving forward, with my school advisor to plan and write my thesis around the political cause and effects of the 1905 Russo-Japanese war! By 1960, my mind and plans were just not into school, a year of thesis writing with other job and family responsibilities just seemed to be too much, but the number of computer courses I took during the program was invaluable to me at the time. However, during the early sixties, I did find unstructured spare time to buy several old homes on Capitol Hill, near the Marine Corps Headquarters, and my father's childhood home on South Carolina St., that I enjoyed restoring and reselling with the help of others Sort of resurrected some of the old architectural instincts still in my head!

Again, at work, there were a number of new developments and changes. Tom Armstrong, an import from the Honeywell Computer Group had been hired to replace Carl Knorr as Vice President, and my immediate head, Dale Holmburg had left the Company to take over a newly formed computer division of Collins Radio in Cedar Rapids, Iowa. He was replaced by Harold Bucher, now my new boss. With the changes, I was now reassigned to the marketing of our newer punch card replacement systems soon to be announced, in addition to continuing the post market support work at the Edgewood Arsenal and the Transportation Centers around the Country. For the next few years, I would be constantly traveling the Country, visiting and living in Bachelor Office Quarters (BOQ's) at virtually every Army base in America! For never being in the Army, I probably have been at, and stayed at, more Army and Air Force bases than military career officers that have served our country. Lee Johnson had been promoted to head the Non-Defense portion of UNIVAC's business, after successfully selling a number of large-scale UNIVAC'S to the Department of Agriculture and the General Services Administration, and shortly within a few more years would be promoted to Vice President of the Government Division, with his eye on the ultimate Presidency of the Company.

Travel was necessary to visit prospective customers for our new line of equipment, soon to be on the market by the end of 1959, a new line of "UNIVAC Solid State 80 and 90" transistorized computers. These computers were one of the first to be announced, replacing the old vacuum tube systems and had both 80 and 90 card compatibility, competing for the mid-level marketplace that IBM had dominated. For the next two years, I was traveling constantly, returning home on weekends. I must have been home more than I remember, because 1960 brought another prideful event in Pat's and my life. That year our second son, Stephen was born at Washington Adventist Hospital in Silver Spring. By 1960 the Army had centralized their evaluation and selection of computers in Virginia, and it was important to influence the Central Command, as well as, the field offices of the Army, as much as we could, in order that their input could be fed into the central competitive bidding process.

It was a tough, ingrained "IBM mindset" throughout the United States Government_and it took an enormous amount of salesmanship and a few miracles, to turn the "tank". And turn it we did, which I will talk about, once I tell you a little on how the industry dramatically changed in the 60s' from "tubes" to "chips".

EARLY YEARS

MIDDLE YEARS

ERWERK
FEUERWERK
FEUERWERK
77

FRANK
WRIGHT
JACK
RAY
LUNCEFOR

ANNE

NEW

YEARS

Please do not feed
ducks / fish

RAY
ANNE JACK

BOOMER
TO SUPER FURY
TO BLACK HOLE
THE TUBE

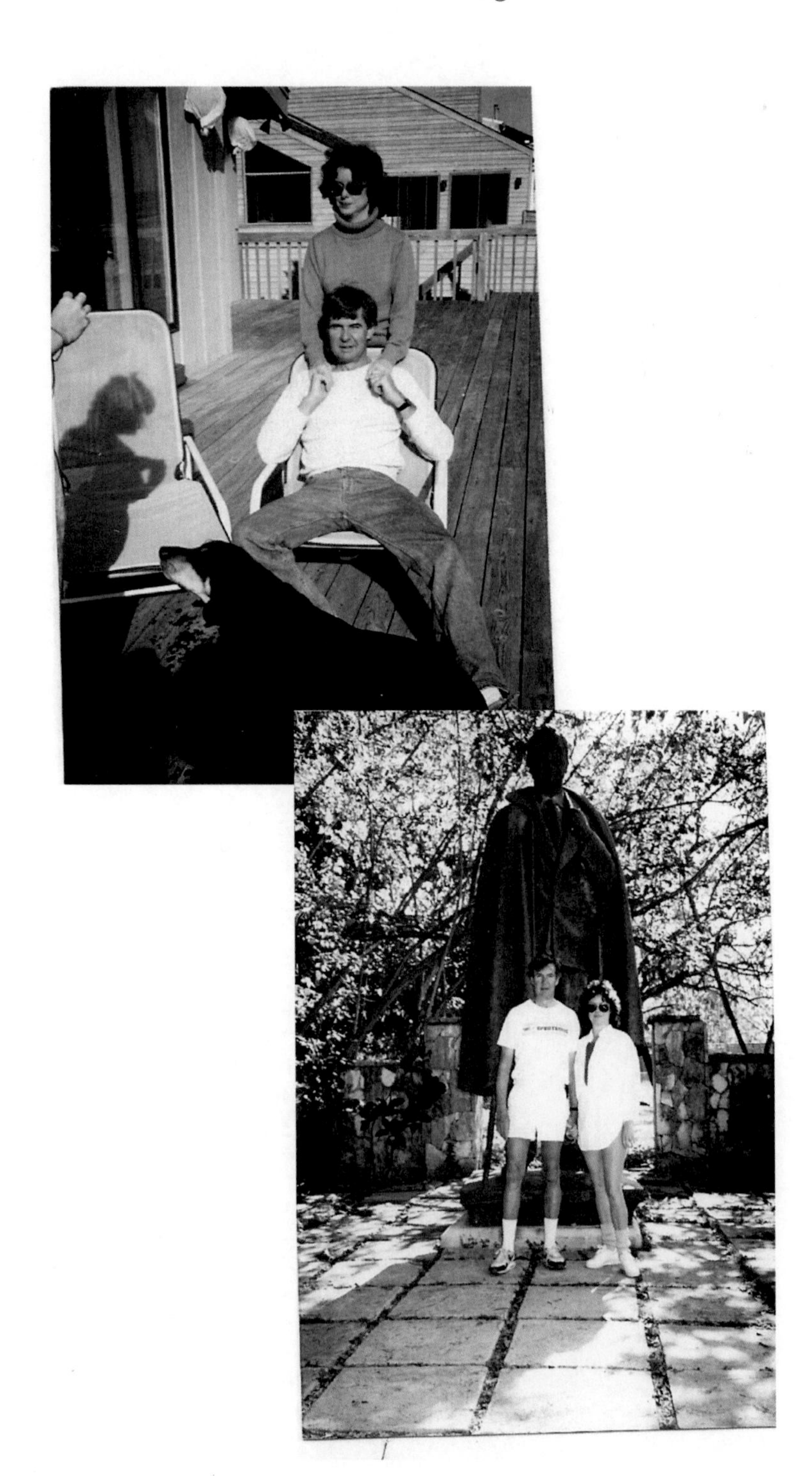

Chapter 8 – "From Tubes to Chips"

By the end of 1960, the era of the vacuum tube computer had been all but replaced by the transistor. The transistor was a device made of semi-conductor materials (germanium and silicon) that can conduct and insulate switch and modulate electronic current. It was co-invented by three scientists, John Bardeen, Walter Bratton, and William Shockley, who worked at Bell Labs in Murray Hills! N.J. In 1951, and changed the world we live in. Shockley later improved on their work and the first transistor was used in a hearing aid built by the Sonotone Company. In 1954, the first transistorized radio the Regency TR1 was built, and I remember carrying around my first portable Emerson transistor radio, about 4x6 inches in size, shortly after. All three were awarded the Nobel Prize for Physics in 1956 for creating the transistor. Shockley was an interesting character and a genius in his field. He left Bell Labs and moved to Mountain View, California to form his own Company to design and build silicon chips using silicon to replace germanium. He was an extremely difficult person to work for and shortly thereafter, a number of his employees left him and formed Fairchild Semiconductor Co., where the silicon transistor was created. From there the industry moved on to integrated circuits and finally the first single bit microprocessor called the 4004 chip, was developed by INTEL and announced in November of 1971. There is always controversy as to "who was first", because Four Phase Systems, a startup company in 1969', formed by ex-Fairchild employees actually built a microprocessor they announced for delivery in 1970 but never reached production until 1972, when the Company was sold to the Motorola Company for over $'250 million, a lot of money in the 70s'! Shockley left the business world and ended up as a Professor of Electronic Engineering at Stanford, where he developed a theory that he called "Dysgenics". Dysgenics preached that all persons with an IQ below 100 should be paid a fee to undergo voluntary sterilization and as part of this theory, he created and supported a sperm bank for geniuses and further claimed that certain races were smarter than others. Certainly an odd ball, but then again, many geniuses are!

In the case of UNIVAC, the last of their vacuum tube machines was the UNIVAC 490, a random drum data storage and retrieval system that was used in both the scientific and business community. From then on, because of the transistor, all systems were made smaller, more compact, and reliable. In addition to size, transistors were less expensive; where the cost of a vacuum tube was close to $20 each, transistors were $1, and tubes which were constantly burning out or causing intermittent failures, would have to be replaced. Transistors also used less power. It was jokingly said that when the old ENIAC, which had thousands of vacuum tubes, was turned on In Philadelphia; it would take five minutes to warm up and it would cause a brownout across the whole City. It is hard to envision the size of these old machines, when practically all you see today are hand held and desk size computers, but if you are ever in Washington, D.C., you should stop and visit the Smithsonian Museum, where they have placed one of the original UNIVAC 1s', which, at one time, was installed at the Census Bureau, where it was used for the first computerized census in the early 50s'. IBM's entry into the solid state world was the IBM 1401, with core memory, and magnetic tapes and drum storage, shortly to be replaced by the advanced IBM 1410. Both of these systems were the most popular at the time, and allowed IBM to continue to dominate and protect their base of old punch card users.

Shortly, thereafter, in 1962, there were two major events that occurred that year. First, my third son Chris was born and that same year, UNIVAC announced the 1004, a highly sought after transistorized punch card computer that was successful in "cracking" the IBM marketplace, and made me one of the top producers in the Company over the next several years. Chris was my lucky charm that year! The 1004 was an extremely fast card processing system that was compatible with using either 80 or 90 cards; plug board programmed and relatively inexpensive, selling for $66,000 or leasing for $1650 a month. Although comparatively small in size, the plugboard was massive, weighing close to 14 pounds, and still had to be programmed wire by wire. Later, the system would be upgraded to include magnetic tape and many large scale users would acquire a 1004 or by then, the upgraded 1005, to use offline to create cards or tapes, thereby saving expensive mainframe time to perform those duties.

Because of this new system becoming available, and our intensive earlier marketing of the Army, the Central Selection and Evaluation Command selected our system in a competitive bid to replace a majority of the army punched card systems worldwide. In one fell stroke we finally had broken IBMs' dominance in this area. Shortly after, a number of the other Defense and Non-Defense agencies followed suit. The same was true in the commercial world, where the 1004 and the 1005 became extremely popular. Over the next several years, we installed hundreds of these machines in the Army, including van mounted systems used in the Vietnam War. Around this time, in 1963, Lee Johnson had returned to our Division from Corporate Headquarters and was appointed Vice-President of the Government Division. Shortly after he returned, Harold Bucher, my immediate boss died and Lee promoted me to run the Army Division of the Company, which by then had grown quite large. So after some 10 years, I finally got around to running something!

Figure 11: Circa 1962: Showing off the UNIVAC 1004

The year 1963 brought something else into our lives. A new house! I was able to trade one of the houses I had renovated on Capitol Hill for a new home in Adelphi, Md., just a few miles from where we were living. The house was a good bit larger than the last, and came just in time, as the family was expanding rapidly; seemed that we added one "kid" for every 1005 I was selling; and speaking of kids, my fourth child, Jacqueline (Jacqui) was born the year after we moved in. Seems like every girl born that year was named Jacqueline, but we didn't care; we had finally gotten a girl!

Chapter 9 – "Beginning of the End"

Those years of 1963-64 were extremely busy. Although we had been adding staff to handle the growing backlog of orders, installing this huge number of orders on a global circuit was indeed difficult and nerve racking. So balancing family and work was constantly upsetting. In addition, Lee Johnson, who had taken over our Division awhile back, and was responsible for my promotion was becoming a difficult person for me and others to work for. Lee was an extremely bright guy, technically astute, marketing sensitive, but when it came to employees, the sensitivity stopped! He was ruthless in his dealings with many, overly demanding In terms of performance, and did much to demean employees in the presence of others. He was especially critical of the few remaining old Rand bookkeeping personnel who had hung on, trying to make the crossover to the computer age. A few did, but most didn't, and it was sad to watch his approach in firing them. One gentleman who worked for me, I wanted to simply meet privately and advise him that he was to be laid off, but Johnson insisted that he spend a week, attempting to learn enough technically about one of our new large scale computers, the UNIVAC 1100 series, and then to make a marketing presentation to him, with Johnson acting as a prospective customer. If he was convincing-he stayed; if not, he was to be fired. Well, the poor guy, rather than quitting, tried; and then in the presence of Johnson, and a few others, he attempted to make a sales presentation that was pathetic! After, he was told that he had "flunked", he pleaded for his job, and then when turned down was fired! It was the most heartless firing session I have ever witnessed in my life. If I did not have a wife and four kids to think about, I would have quit the moment he was fired. But stay I did, and along with others, bore the brunt of his abuse for a little less than two more years. From time to time, over the years, I think back trying to understand the man and his methods. He was a small person in stature; was struck by polio, as a kid, that left him with one leg shorter than the other, so that he walked with a slight limp; and was extremely near sighted. He was a workaholic, and it was not unusual for him to work into the early morning hours, forcing other employees with families to do the same, or minimally calling an employee at home after midnight to discuss an issue. It was not unusual at all for him to call all managers in for an all-day meeting on Saturday or Sundays, or even midnight on a week night and publicly insult them in the presence of others; It seemed to me, at the time, although others that were there disagree, that he found pleasure in attacking the groups who were not successfully marketing large scale computers, which included mine; always threatening, never supporting!

Time passed quickly and we continued our success in selling additional systems to the Army, but by the end of 1965, I began thinking for the first time, of leaving the Company. During that year, I was approached by Dale Holmburg, my old Manager, who had left, several years back, and was now President of Collins Radio Computer Division. He invited me out to Cedar Rapids, for a job interview with the Company; so taking a few days of vacation, I flew out to Iowa and met with both Dale and Arthur Collins, the founder and President of Collins Radio, who offered me the position of Vice President of the Computer Division. Collins was a late comer to the industry, and with only one product to sell at the time, I decided to decline the offer; since the offer meant moving the family to Cedar Rapids, and after touring the City I knew that Pat, as well as myself, would not be happy in Iowa.

I should probably stop here and mention briefly, a little on my "first start up" Company. While still working at UNIVAC, in 1965, I was approached by a friend of mine, Jim Trebach, who was married to Janet, an old classmate of Pat's, about an entertainment business that was enjoying great success out West and was slowly making its way across the Country. It was an indoor toy sized car racing sport called "slot car racing" that kids and adults were crazy about. Customers bought cars from the store owner and then rented racing time on the store tracks. Jim had a golfing partner, George Vincent, who had started one in the area that was wildly successful, returning one's investment In as little as six months! It seemed like a great idea, so I agreed to start the Company, as President, and raise $150,000 to fund the first two centers. Overnight, Model Car Management Corp. (MCMC) was created, and at night, I developed a business plan to raise the venture capital, and after a number of night time presentations to prospective investors, we raised the money, but raising it was not easy. Most of the investors were friends or family members; folks like my dentist, a few computer associates, and one individual I'll never forget. In my youth, when I lived in Brookland, I used to have my hair cut by a barber named Dan Stallone, who was the Uncle of "Rocky" Sylvester Stallone, and I convinced him at one of our presentations that he would "make a fortune" if he invested in slot cars. Dan, who was an extremely conservative guy, finally decided to invest $1000 in the venture. That investment haunted me all thru 1966, and I'll tell you why in a minute. We came up short of the $150,000 needed, but each of the Officers, took out loans and raised the balance. Armed with investment money, we leased two large buildings that I had to personally sign for; and not knowing the potential implications did so! We shortly were up and running and the two stores were "packed" from the start. Kids and Adults both, buying scores of racing cars and renting enormous amounts of track time. Then, all of a sudden, it stopped! It turned out that after spending, in some cases, up to $1,000 on different race cars, parents would tell their kids; no more! Enough is enough! And so we folded the Company and went bankrupt! Stallone was livid, and for a while, I looked around every street corner, thinking he may be sending "Rocky" after me. Our landlord chased me for the balance of lease monies owed; vendors were demanding monies owed; and I personally ended up deep in debt, even as I negotiated myself out of the building leases, with an extremely "kind" landlord, who lectured me in the ways of business, before he let me "off the hook"! Later, in 1966, I personally paid Stallone back his investment money; I felt so bad about his investment, since it was a good part of his savings. However I never went back to his barber shop for a shave! At any rate, MCMC taught me a lot about the real business world, and the lessons learned in that venture were going to be of much value to me in future years.

That was not the only disaster that happened to me during that period of time. Between the stress of work and the bankruptcy of the "slot car" company, one night while driving back from southern Maryland on Route 301, my Sunbeam Alpine convertible was hit from behind. I suffered a badly broken arm, and a fractured skull, while the car was totally demolished. I was immediately helicoptered to Holy Cross, a local hospital, where I underwent immediate surgery, and spent a couple of months out of work. At the time I didn't think anything worse could have happened but as the years went by, hard to imagine, a lot more tragedy was in store for us. That summer more bad luck came our way, when my youngest son, Chris, just a toddler, overturned a gas can in the lower part of our home, near the furnace room, turning the house into a raging inferno! Thank God Pat had smelled the smoke fumes before the fire raged and was able to get all of the children out of the house before it was largely demolished. I was at work, at the time, in D. C. and rushed home just in time to see the firemen extinguishing the last bit of flame. Thankfully, Ray Lunceford and I had bought two investment houses at an auction, earlier that year; and with no place else to go, I moved the family to one of the vacant homes until we could rebuild. Where would it all end?

When, I returned to work that fall, I felt that my career was over at UNIVAC. Pressures on me mounted over the rest of the year, and in a last ditch of spitefulness on the part of Johnson, I was sent to St. Paul, Minnesota, our Company Headquarters, to attend a training class on large scale computers, the week before Christmas, with outside temperatures hovering at 32 degrees below 0. Shortly, after the Holidays, I was summoned to Lee's office and informed that it was time to go, and sadly, with close to 12 years with the Company; during the most fantastic period of its industry growth, I left. Fired by a little Napoleon for" absolute cause"! If I would have been asked to give him a sales presentation on large scale computers to" save my job", I would have won "hands down", but that would have been pure fantasy on my part, to think that would happen. In the end, the firing would turn out to be the shot that turned my life around and hurled me into a more exciting career orbit.

Chapter 10 – "Software Landing"

Within a week after leaving UNIVAC, I had calls from three different Companies that were interested in meeting with me to discuss joining their companies; Computer Sciences, Aries Corp, and the third, Federal Data Corp., a Company I had never heard of. I knew both Bud King, now with Aries Corporation, and Vince Grillo, with CSC, since both had worked with me at UNIVAC. So, first Intrigued by the call from Federal Data Corporation, I called back and met with Bob Hanley, the President, shortly after, and he described his Company and his interest in me. It turned out that the Company was a startup, and I do mean a startup! The Company consisted of one person, himself, and one other angel investor, sharing a one room office, on Wisconsin Avenue, one block from the newly reopened UNIVAC offices, in D.C. Federal was starting off as a computer leasing Company and they were interested in buying the lease portfolio package of the several hundred Army 1004 and 1005 UNIVACs' that my Army group had leased and installed, and felt I would be the best person to hire, since I had personal contacts with the customers and the equipment. I knew that UNIVAC was interested in selling a variety of leased systems, but to a one man startup! He had to be kidding! The meeting ended with Hanley offering me a 5 percent "stock option" in his new Company, and a low salary. We shook hands; I told him I'd think about it and get back. A 5 percent stock option in a “startup"; with no apparent business, and me having an instant need to support a family-- no way! I could have started a new race car business and own 100 percent of nothing, and be the "President" besides. The next day I interviewed with Vince Grillo by phone in Los Angeles, who offered me a job as Director of Corporate Marketing, initially working out of Washington, a $35,000 salary, plus bonus and 5,000 shares of CSC stock, to set up a National Marketing organization. At the time, CSC had annual sales of around $30 million; still a small Company in a new world--Companies that sold software, not hardware! You see, back in 1965-66, software companies were few and far between, Companies such as CSC, Computer Usage Co.,

System Development Corp (SDC), Informatics, Aries, and several others were the only significant Companies around. Most consistently run by technical types, who by nature distrusted marketing talent. Since customers were always looking for software and systems support help, it was not difficult to sell your capabilities, customers came looking for you. Times were beginning to change in the mid-sixties, however, when competitive software companies began to spring up and challenge each other. All of a sudden, a technical individual needed marketing support in order to find and close prospective business. So, a new industry had suddenly erupted! I thanked Vince for his interest in me, and the job offer, telling him I would get back to him shortly after talking it over with Pat. I knew there would be a lot of travel with the position, and I needed to know if she would be comfortable with that. I next called Bud King and met with both he and Dick Daly, a few days later, and was offered a position as Vice President of Corporate Marketing at the same salary and 25,000 shares of Aries stock, but no assurances as to whether they would entertain a large expansion of the marketing operation right away. I had had a good working relationship with both at UNIVAC, and after mulling over both job offers, I decided to accept the Aries offer, mainly because there would be little travel, which appealed to both Pat and myself, So, the next Monday morning, after calling CSC, declining Vince's offer, I showed up at Aries for work; met with Dick and Bud, and then had "get together" introductory sessions with various technical groups. By 2:00 that afternoon, I walked into Bud's office and amicably resigned! The technical folks were cool, and some even hostile, toward my joining and directing their sales efforts. It was obvious that it would be an extremely difficult process to win them over and grow the business. Next came a hurried call back to CSC, and shortly I walked into their front door in Maryland, reporting to work. CSC had been started by two individuals, Fletcher Jones, and Roy Nutt. In 1959, with $100 and by the time I had joined them, it was the largest software company in the world. Fletcher was the flamboyant of the two; the leader; the salesman, whereas Nutt was an extremely talented software developer, who designed a number of sophisticated software compilers for various companies, including a FORTRAN product for the UNIVAC large scale computers. He was a" nerdy" type, very shy guy, who drove a

Volkswagen "Beatle" and talked software! I first met him in St. Paul, when he was hired by UNIVAC to teach the class I attended, the week before that past Christmas. Fletcher, was just the opposite, a dynamic marketer, outgoing individual; married, divorced and dated one of the. “Miss America" winners in the early 60s', drove the "hottest" and latest Rolls Royce’s on the road; and was killed in 1972, when his Bonanza jet crashed into the side of a hill while flying to his home in Beverly Hills, from his farm in San Yanez, California, where he bred and raced thoroughbreds; and collected priceless pieces of art. Dead at the young age of 42! Sometimes, when I'm reflecting on my own career; my interest in art and horse racing, I think of Fletcher, and how he may have been a heavy influence on me when I also started investing in the thoroughbred industry back in the early 80s'. When I first joined CSC, I got to meet and sit in a number of marketing meetings with Fletcher and Vince, where I had the uneasy feeling that I was a minor league player, in the presence of two major leaguers. They were significant players in our technology industry, as was Bill Hoover, who replaced Fletcher as Chairman and CEO in 1972, after Fletcher had been killed. Bill had joined the Company in 1964, from NASA JPL, where he was Director of Computer Operations, and became President of CSC, in 1970, when Fletcher relinquished the Presidency, staying on as Chairman. During my time at CSC, I learned a lot a lot from both Hoover and Grillo traveling with them, making sales calls together, planning/strategy meetings and developing overall business plans. As I mentioned, earlier on, I was one of 9 employees, including Hoover and Grillo, which developed the business plan and strategy for the Company entering the time sharing service business.

The remarkable thing about CSC was that whereas the West Coast headquarters was extremely marketing driven, the East Coast Operations, where I was domiciled, was just the opposite, with suspicious, coolness toward marketing types, much like the Aries operation! So armed with the backing of Hoover and Grillo, I began the chore of building the very first formal marketing division for the Company, except for Hoover and Grillo, who at that time were the entire sales organization of the Company. I still remember Hoover's comment to me, one day, when we walking the halls of the Pentagon toward a meeting, when he stopped and said, "Jack, this is a fantastic building, every doorway is a marketing opportunity". As it turned out, it was and the Company went on to win a number of contracts, with major Defense Department Agencies, through the efforts of a number of talented marketing people I hired. However, there still existed an aloofness from the Technical Operational people, who still kept their distance from marketing, and yet it was essential to involve them in the sales process in order to influence and assure our customers we could satisfy their needs. A massive amount of my time was spent pacifying this group, pulling them into the process, while at the same time producing the necessary quota performances dictated and commanded by the West Coast Headquarters. When Hoover would visit the East Coast, it seemed that the entire East Coast Operations became "marketers"-- and when he left the "divisor" wall would reappear! But over time, remarkably things changed, and as we started winning additional multi-million dollar contracts, barriers started coming down, and we "all became one". When I left the Company in 1968, it had sales of well over $65 million and by 1998 over $16 billion, thanks to the combined marketing and technical efforts. Today the Company is still a multibillion dollar Company with a large international sales organization, much, much larger than what I started there, but I am proud of the fact that what I started and began, contributed in a big way to today's size and reputation, that the Company enjoys.

Today, Aries Corporation is gone! A victim of more marketing driven Companies, but interesting, however, Federal Data Corporation survived and grew into a large Company, that years later, became a formidable competitor of C3, (my Company), and at one time, shortly after C3 went public, we came close to acquiring Federal Data for almost $50 million. When our offer failed, Hanley later sold his Company to the Carlyle Group and retired to Florida, a very wealthy man. Hanley had been blessed by picking out a talented group of people, and had the vision to hire, and promote Dan Young as President, a seasoned and talented computer executive, and Dan stayed at the helm of Federal Data for over 25 years, growing it to one of the more formidable system integrators on the East Coast, with sales well over $100 million when sold to Carlyle. Dan has now retired and spends his time between his homes in Virginia and Florida, but still keeps active, sitting on several Technology Company Boards; and a person I consider a good friend, even though we competed on many major contracts over past years. He was a real "pro" in the Technology Industry and the Industry will miss his many talents.

In the less than two years, I stayed at CSC, Pat and I met and made friends with a number of employees that I occasionally still see today, including a bunch of the technical types that disliked marketing people, but little by little, I had worn them down, and "Peace pacts" were made. Men like Paul Kazek, Vice President of the scientific Division, Jim Trawick, Director of Business Systems, Robert Head, the creator of one of the first Management Information Systems (MIS) developed by anyone. Little things, you remember, Bob lived on a little farm in Virginia, raised chickens and his wife sold eggs to the office staff, that Bob would bring in. I don't know how that goes on your resume! Fred Henschel was another friend I met at CSC. He had been hired by the company, as Director of Personnel, hiring analysts, programmers, marketing people and whoever else we needed to provide talent to our expanding contract base. Fred has been a lifetime friend, and a part of my last Company, Computer Equity Corporation (Compec). He left CSC around 1968, to move to New York where he worked for Saul Steinberg, a major player in the computer leasing and reinsurance industries in the 60s' and 70s', as Saul's Administrative Assistant, Fred was involved in all aspects of Saul's far flung empire, and left a number of years later to assume the Presidency of the National Bank of Washington, in D.C. On a minor note, I wanted to mention that Saul's son, whom no one ever hears about, started a financial newsletter, years ago that folded, but he ended up marrying Maria Barteloma of MNSBC notoriety in one of the largest and expensive society weddings in the "Big City". Curiously, there are seldom articles or statements in the news, when it comes to the couple. Saul, Jr. seems to have dropped thru the woodwork, and I am not sure they are still connected at this point in time. Sadly, Saul, Sr. had a massive stroke in the 90s' and is wheel chair bound, and selling off major parts of his business empire.

Going back to CSC for a bit more, the stress and business travel, I spent at UNIVAC ended up being not much different. The almost two years I spent with the Company, was moving from Office to Office; helping to setup marketing operations, both commercial and government, in CSC key locations throughout the

Country, mixed with marketing and planning meetings in El Segundo, California, the Company Headquarters. The constant meetings in El Segundo were probably the main reason I finally had to leave the Company. Hoover and Grillo had a "little bit of Lee Johnson" in them, in that they would call impromptu meetings in California from time to time over Saturday and Sunday weekends. It meant that I would have to catch a late afternoon plane to Los Angeles on a Friday and catch a "red eye" flight back to D.C. Late Sunday night. Working weekends was not the problem. In the business world, they are part of the equation of success, and over the years, I put in many of those night times and weekends to make things happen. In the case of CSC, the weekends were in L.A. 3000 miles away from my family and marriage, and I no longer wished to rupture those relationships. It was time to go! My resignation came to a head in the middle of a meeting in L.A. with Hoover and several members of my marketing group, on a Sunday afternoon; I looked at my watch and noticed that I still had time to catch a one o'clock plane back home, thereby skipping the red eye later midnight flight. After a split second thought, I rose in my seat and announced that I was resigning. There was shock, in the room, with Grillo and Hoover walking with me, on my way out, attempting to discourage me from leaving. I caught the early plane, just in time, and once more I was --UNEMPLOYED!

Chapter 11 – Computer Data Systems "Next Stop"

I should start this part of my story to tell you, that my departure from CSC was not as hasty and abrupt as I may have portrayed it in the previous paragraph. Actually, during the early part of 1968, I had been entertaining the idea of leaving and starting my own software Company. This industry was still in its infancy, and the time spent at CSC gained me a tremendous amount of knowledge in the software world, and the expertise to develop a well-rounded business plan to attract investors and successfully compete. So, in my spare time at night and on weekends, I developed a workable plan, along with a fellow CSC employee, Jim Trawick, an ex-IBM'er and a key technical Director at CSC, who was planning to leave and join me, as a co- founder of the Company, which we named Computer Data Systems, Inc., or, CDSI and incorporated the Company on July 18, 1968. As it turned out, Jim, at the last minute, "got cold feet" and stated, "I just can't take a chance, Jack, I've got to back out"! At that time, after unsuccessfully trying to change his mind, I wished him well and, within two weeks, convinced an old acquaintance, from my UNIVAC days, Joe Easley to join me. Joe, at the time was working for a small Minneapolis software company as their Eastern Regional Manager, and had a solid technical background, and interesting enough, a flair for marketing as well. We then set about trying to raise in the neighborhood of $200,000 through friends and business associates, and by the end of August we had done so. Our biggest investor was an old friend of mine, Tom McCaffrey, who had moved to Atlanta as the Branch Manager of UNIVAC. I still recall my meeting with him, in my Atlanta Hotel, where after almost four hours of conversation, I convinced him to invest $50,000, in the stock of newly started CDSI! At the same time, I began to talk to several key technical and financial people that would be needed, once we had obtained adequate financing. Walt Scott and E.J. Collins were two outstanding technical employees at UNIVAC, who were eager to join, whereas Clifford Kendall, a friend from College days with a finance background, was interested, but would not agree to join until I had obtained enough capital to guarantee him a year's salary.

Kendall, at the time was working in the mid-west for Booze Allen, consulting, and supporting Booze Allen clients. Shortly, I leased a small apartment in Downtown Silver Spring on Colesville Road, and setup offices for the four of us with Easley, as Executive VP, Collins, as V.P. of Technical Operations, Scott, as technical support and myself, as President, CEO. Shortly later, Kendall agreed, once I assured him monies had been raised, resigned his job and joined us as the Company's first Vice President of Finance.

Soon, we were able to capture a few small contracts to establish a revenue base, and add a few additional personnel, so that by the end of 1968,, we had grown by more employees, that would ultimately set the stage for the Company to grow over the years to over 3300 people and revenues north of $300 million, when the Company would be sold to Affiliated Services Corporation for $373 million in 1997.

B-8 Business, Stocks THE EVENING STAR Washington, D. C. Friday, November 22, 1968

Of People . . .

Computer Firm Names Officers

Figure 12: CDSI Founding Fathers

During my time at CDSI, from founding the Company in 1968 and leaving in 1970, a number of significant events happened; some because of my efforts, and others as the result of Joe Easley's. Although we were still an extremely small company, we had need for additional capital that. I could not raise privately. It was just by sheer luck, that my friend Ray Lunceford had met a new friend, a few years ago, while serving in Korea, by the name of Bob Ryan, a New Yorker, with a top management job at one of the New York Merrill Lynch offices. I met with Ryan, thru Lunceford, on a trip to New York and he in turn said "you got to meet Charlie Plohn-he can help you"! Next was a meeting with Charles Plohn, at his office in the Pan Am building in New York, The meeting was indescribable! One would have to have been at that meeting to personally experience it. Charlie loved nudes, so all thru the various office, walls were hung with elegant expensive paintings of nudes; some erotic; some not. Charlie had a very simple command of the English language with almost every word he uttered ending in four letters! The man couldn't complete a sentence without foul language being a part of it, but he was brilliant at what he did. Plohn had built a very successful investment business, by taking small companies public, selling $1 million IPOs', in rapid fashion. Forbes magazine had written a feature story in 1967, about Plohn & company, calling him "One a Day Charlie", referring to the rapidity with which he took companies public. After Ryan introduced me to him, the meeting lasted less than one hour, at which time he said "How much do you need kid", and before I could get into a "need" discussion, he literally writes a $1 million IPO offer on a piece of note paper agreeing to sell 250,000 shares at $5, hands it to me, and tells me to go talk to Bob Arum, his general counsel, who at that time worked for the Louie Nizer law firm, and was responsible for all of Plohn's legal investment banking work. So I went to see Arum; signed some papers; and the deal was done; I was rich - owned 66,000 shares of the stock worth over $300,000 on paper after the offering; and had an employment contract paying me $26,000 a year. Some of you readers may have heard of Bob; he soon was to leave his law firm, and start Top Rank, Inc. taking with him a young attorney, Mike Heightner who also worked on our account for Plohn. Arum was a tough, brash and bright New York attorney who loved boxing and was an avid fan of the sport, and as it turned out on one of my trips to Charlie's office, I was in his reception area sitting across from another IPO candidate who turned out to be; and you won't believe this, “Cassius Clay" later to be known as Muhammed Ali! I was briefly introduced to him, and found out later that he was there, thru Arum, to raise IPO monies for a new Company he was fronting called "Champburger", Champburger was a startup. McDonald's had opened one fast food chain store in Miami, and Clay was to be the "pitchman". End of the story-Champburger went bust and Arum and Ali went on to fame and fortune in the boxing arena, competing with Don King; and Computer Data Systems, now a public company, listed on the NASDAQ, under the symbol CDSI, went on to do the same in the year 1968, with a new war chest provided by Plohn.

PROSPECTUS

250,000 Shares of Common Stock

COMPUTER DATA SYSTEMS, INC.

(PAR VALUE 10¢)

THE SHARES OFFERED HEREBY INVOLVE A HIGH DEGREE OF RISK. SEE "INTRODUCTORY STATEMENT".

At the present time there has been no market for the Common Stock of the Company. The price at which shares are being offered hereby was determined solely by negotiation between the Company and the Underwriter.

THESE SECURITIES HAVE NOT BEEN APPROVED OR DISAPPROVED BY THE SECURITIES AND EXCHANGE COMMISSION NOR HAS THE COMMISSION PASSED UPON THE ACCURACY OR ADEQUACY OF THIS PROSPECTUS. ANY REPRESENTATION TO THE CONTRARY IS A CRIMINAL OFFENSE.

	Price to Public	Underwriter's Discounts, Commissions and Expenses (1)(3)	Proceeds to Company(2)
Per Share	$5.00	$.55	$4.45
Total	$1,250,000	$138,000	$1,112,000

(1) Includes $125,000 ($.50 per share) cash commission and $13,000 ($.052 per share) cash for expenses of the Underwriter and its counsel fees partly non-accountable.

(2) Does not include additional filing, printing, legal, accounting and miscellaneous expenses of approximately $37,600 ($.15 per share) which the Company must pay in connection with this offering.

(3) Does not include substantial additional compensation to be received by the Underwriter. See below.

This offering involves:

(a) Special risks concerning the Company. For information concerning such risks see "Introductory Statement" page 3.

(b) Immediate substantial dilution of the book value of the stock from the public offering price. For information concerning such dilution see page 4.

(c) Significant additional underwriting compensation through the purchase by the Underwriter, and associates of 25,000 shares at $.40 per share shortly before this offering and other concessions involving preferential rights to future financings, rights to designate a nominee to the Board and indemnification. The underwriter had purchased an additional 10,000 shares at $.40 per share. On March 7, 1969, these shares were resold at $.40 per share to officers of the Company because of objections which were raised under the policies of the National Association of Securities Dealers as to the fairness of the underwriter's compensation. For specific information concerning these factors, see "Underwriting" page 12.

At the Company's request the Underwriter has reserved, until the day following the date hereof, 60,000 shares of the Common Stock being offered by this prospectus for sale at the public offering price to officers, directors and employees of the Company, friends and relatives of such persons, and persons who have indicated an interest in the affairs of the Company. Accordingly, the number of shares available for offering to the general public will be reduced by the number of shares, if any, purchased by such persons.

CHARLES PLOHN & CO.

The date of this Prospectus is March 18, 1969.

Figure 13: CDSI Prospectus

Arum made several trips to our offices the beginning of 1969, to review our progress, often accompanied by Heighnter; offered suggestions, and quickly disappeared. About the only solid advice Arum offered me on one visit was "Jack, get rid of that "hay seed "attorney, Steve Leonard, you have - he's a loser-I'll find you someone that can really help". Looking back, in time, his advice I wished I had taken, but more later on that story. Plohn shared a

Board seat for a short time, but never attended meetings- too busy doing "one-a-day deals" I suppose. We had earlier expanded our board to include our attorney Stephen Leonard, D.C. via of Indiana, and Dr. Ray Hoxeng, President at that time of the Inter-American University in Puerto Rico, a friend and client of Leonard. Those two, along with Easley, Plohn and myself, made up the Board, at the time of our public offering. Later, a chap by the name of Peter Neumann, a Vice President at Bache & Co., now a long gone brokerage house, that I had befriended, also joined the Board, after the public offering. The following year, on June 26, 1969, through Bob Ryan's connections, I was able to get CDSI listed on the National Stock Exchange, with a new Exchange symbol COD that it maintained until the Company was sold many years later. The National Stock Exchange originally was the old Commodity exchange in Chicago that had branched out and added Companies like ours to their listings. Ultimately the National Exchange moved to Boston, and became the Boston Stock Exchange. To the best of my knowledge, Computer Data was the first Washington Company to be listed on the Boston Stock Exchange.

Earlier, in April of 1968, I had convinced one of my CSC customers, a young man by the name of Jim White to join the company in a marketing position, and Jim, who was from Florida, was closely connected to the State of Florida through a FSU college classmate, then working for Governor Claude Kirk. It was through Jim's connection, that I got to meet a number of top State Government data processing people, in Tallahassee, and since Jim's friend was Governor Kirk's Assistant and Chief advisor, I was introduced to Kirk, and we connected immediately and shortly thereafter received a major contract with Florida to develop a management information system for their Dept. of Education. On several occasions, Jim and I would fly back to Washington, at Kirk's invitation, on the State's Lear Jet, where he conducted various State meetings with different Federal Agencies. He was an outgoing, charismatic personality, who thought "big" and married "big"! I remember well, one flight back to Washington, with Jim White, Kirk's Administrative Assistant, myself, and Kirk, along with Kirk's third wife, Erika Mattfeld, a blonde, buxom German born actress, he had met shortly before he had taken office, during a failed business venture in Brazil. Those two, were one of the most colorful couples

I have ever met. That trip was a blast with Kirk regaling us with stories of his past and present, such as riding a horse to a news conference and planting the state flag on the ocean floor, vowing to use state-owned airplanes to defend Florida's territorial rights. He vetoed 48 bills during his first year in office and newspapers dubbed him "Claudius Maximus"! His favorite saying was "The garden of controversy must be continually cultivated", he told Time Magazine back in 1967, "otherwise nobody knows you are alive". When he was being criticized by news outlets for being an outrageous huckster, Governor, Kirk replied: "I'm just selling orange juice, more orange juice, selling Kirk, selling Florida, people are paying attention"! That contract we received, we got, because of him, and it launched us into opening a new Southern Regional Office in Tallahassee, hiring a young man, Lee Rathbun, to manage it. Fond memories of exciting times with Kirk and his staff members will always remain with me. He was a character to the end; three marriages; various business ventures; ran for several political offices and helped Lee Iacocca, after Ford Motor fired him; get the top job at Chrysler through his auto industry connections. Just a one of a kind personality that comes along every century!

By October, 1968, using my earlier connections at UNIVAC and CSC, I was able to capture a few commercial contracts along with a number of contracts in the Government Arena, specifically one with the U. S. Navy, that I obtained thru one of my ex CSC customers, Carl Calo, a friend and high level Navy Government employee, who gave me one of my first breaks; a contract that provided a stable base of revenue and became a mainstay contract for the Company for years after I left the Company. Other contracts were obtained through UNIVAC, such as the National Dry Cleaners Association, and the Machinist Union, gained thru Tom Flor, an old friend from UNIVAC, who introduced me to his customers. Tom happened to have played Football for Jim Tatum at Maryland, when I was there, and he was a tremendous help in providing me with other business contacts later on. Other contracts followed quickly. A NASA consulting contract won through Dr. Birbaum an old friend at NASA, from my days at CSC; a significant contract with Air Alitalia, on a visit to Rome, that I obtained through Tom McCaffrey who had earlier helped finance our company. Tom had been promoted overseas by UNIVAC to Vice President of Southern

European Operations, and provided me with a sub- contract that he had with the Airline. Through several connections I had developed, previously with top level officials in the Department of Commerce, I was introduced to Rocco C. Siliciano, who at the time was the Secretary of Commerce. Siliciano was instrumental in staging trade mission shows to various parts of Europe and Asia promoting the sale of American products overseas, and felt our small Company would be an ideal Company to be a part of a technology trade mission to Western Germany that his people were organizing. So, we were selected, and I joined five other Companies, including Motorola, and Control Data Corporation, to participate in the Mission in May of 1969.With the assistance of Department Of Commerce personnel, we toured major West German Cities, meeting with, and signing business agreements with several German hardware and software Companies, such as Nixdorf Computers and several smaller software and hardware peripheral companies.

Figure 14: With Secretary of Commerce, Rocco Siliciano at German Trade Mission Sendoff Meeting

Visits were made to Munich, Hamburg, Padderborn, Düsseldorf and West Berlin, culminating in West Berlin at the American Embassy, where we were given a large farewell dinner and party hosted by the American Ambassador to west Germany (see opposite page picture). It was here that I met two fellow trade show members of our group, Mort Wimpie with CIT Leasing and Joe Mahoney, with Control Data Corporation, who would become lifelong friends. Mahoney was a handsome, extremely articulate Irishman, who had spent quite a bit of his computer life in Europe, traveling back and forth from the U.S. selling computers to European companies, and it was during this period of his life, when he became a personal friend of Prince Rainier, the reigning head of Monaco, at the time, along with "Friar" Tuck, the Prince's confidant assistant and traveling companion. Mahoney, became close over time with Ranier, prior to the Prince marrying Grace Kelly, and also travelled with him, and visited him often at his palace in Monaco. I heard many stories from Joe, over the years, about their trips together; the trials and tribulations of some of the Prince's personal life, and the initial reluctance of Rainier's upcoming marriage to Grace Kelly, whom he would ultimately marry after with Friar Tuck convincing him to marry. Joe , along with Friar Tuck, actually convinced her to marry Grace, on a previous bachelor's trip to a European mountain trip. Joe met Grace on a number of occasions at the Palace, and attended the Royal Wedding, in Monaco, along with his first wife. Afterwards, Joe was often invited to visit the couple after their marriage and spent many a night with both the Prince and now Princess Grace over dinners regaling Grace, with their stored "bachelor" times together. Joe's first wife, who I knew well, unfortunately died a long and painful death from cancer, while they lived in the Washington, D. C. Area, and shortly after her death, Mahoney moved west to Hillsborough, California and later married Lilly, a beautiful woman that he had met in Denmark. Today, they both are still residing in Hillsborough, California; Joe retired and is still painting beautiful watercolors of the orchids they both raise; and Lily, his wife, still working as an oncology nurse in San Francisco. For many years after, they both were often invited to spend time with Grace and Rainier, and did, both in the States and in Monaco, until Grace was killed in that well publicized auto accident. After then, Joe has told me he seldom saw Rainier afterwards, although they did correspond, from time to time. Anne and I have visited them in California, several times, but most of their traveling these days is spent traveling to Denmark visiting Lily's family; with the

The Evening Star

WASHINGTON, D. C., FRIDAY, JUNE 27, 1969

WASHINGTON FIRM ON NATIONAL

Computer Data Systems, Inc., Silver Spring, began trading yesterday on the National Stock Exchange. Opening price was $6 per share on the company's 250,000 outstanding shares. Company president John C. Bailenger, left, said CDSI raised $1.25 million when it went public over-the-counter in Maryland in March. At right is Ted McCormack, president of the exchange.

Figure 15: Listing on National Stock Exchange

balance of their lives cultivating their ever expanding orchid garden. Concurrent with my marketing efforts, Joe Easley was also making connections with his stable of contacts and obtained a contract with Shell Oil in Paris, France, as well as a study contract in Finland, with Kesko-Oy, a wholesale food company. Walter Scott, one of our now four technical support people was sent to Finland for several months to handle the contract support, and Easley began commuting back and forth from our Corporate offices to coordinate the work of a part time staff we had hired to complete the Shell-Oil and Air Alitalia contracts. More and more, Easley was spending his time commuting back and forth from our Maryland Offices and Europe, running up excessive expenses on contracts, not creating profits that we needed. This had begun to cause a running conflict between the two of us that also began to cause some consternation amongst several board members, fueled by critical commentary from Kendall.

Figure 16: First CDSI Shareholder Meeting 1968.

However, we continued to grow our business base, but because we were still basically in a startup phase, our first public annual report for the year ending June 30, 1969, showed revenues of $281,937 and losses of $127,651, but a decent balance sheet depicting over $1,250,000 in net assets. For still a small company, with our foreign subsidiary, CDSIL, shortly to receive an additional cash infusion of another $1,250,000, it was not really a bad startup, when you compare it with many startups today, or some of those back in the “dot com" days.

A new lucky break came for us on a second U.S. Trade Mission to Europe, that I was able to arrange through the success of the first. I elected to send Joe Easley, as a substitute for me and that particular mission turned out to be highly successful, due primarily to the efforts of Easley. While on this U.S. Trade mission show in Europe, Easley made a contact with Arthur Humphries, then Managing

Director of International Computers Ltd., in London, one of the largest European computer manufacturers, at the time, who had an interest in our Company and met with Joe to discuss our interest in marketing some of their large scale computers in the States, But after a few meetings, he took a special liking to Easley, and before long, based on Easley's persistence, Humphries agreed with us to establish a new European subsidiary; establishing a European Time Sharing business called Computer Data Systems, International (CDSIL), headquartered in Paris, at the old UCC time sharing center, and a second time sharing in Geneva, Switzerland. ICL was the leading competitor to IBM in Europe and had decided to form this partnership with us by buying 25% of CDSIL for $1,250,000 in cash with the intention of committing against IBM in the European time sharing business. A real coup for us. We had raised almost the same amount of money in this subsidiary company that we had just raised a year or so ago for the parent Corporation, increasing the assets of the parent dramatically, without dilution, and by the end of 1970, we had approximately 100 employees around the Globe.

Easley then expressed an interest in moving to Paris to run the entire overseas operation; agreements were reached and he moved early that year with his family, to Europe, with my blessings, since by this time, the two of us were having major personality problems, based on his disregard for financial matters. Joe was a superlative marketer, with deep technical skills, but I had many problems with him in the States, and later in Europe, regarding his spending habits. He was a terrible P&L manager; spent monies foolishly, and it was not long before our total European venture crashed, a piece at a time.

But well before the end of 1970, because of the clashes between myself and Easley, the two co-founders of the Company, and the concerns of several board members, Easley moved back to the States and resigned; moving back to his hometown in Oklahoma, opening up of all things, a motorcycle franchise. By now, I also had come under increasing pressure from the Board to more swiftly turn the Company around to a profitable state, with Kendall siding with Board Members, attempting to convince them that our European Operations under Easley were a mess, which they were, and that our Government business, under me, was being bid and obtained at non-profitable rates, and therefore unless changed would destroy the Company. At this stage of my career, I was following the same bidding and pricing strategies learned at CSC and UNIVAC, that ultimately would have resulted in the same profit results that these firms enjoyed, but the CDSI board, led by Leonard, a neophyte, in terms of contract bidding, as were the other board members, began listening more and more to Kendall, whose background in our business contract base was also minimal, limited from his short time with CDSI from the day I hired him. One of my closest college friends, and a person I had recruited and gave a substantial stock position and corporate position to, had turned against me!

Shortly after, I called for a board meeting to be held at our office in Paris, France, to finish the planning and closure of our overseas operations, with the plan to focus back home on the commercial and government business. As it turned out, I had decided to take my wife, Pat, and our oldest son, Glenn, with me to Paris, so that they could enjoy a few vacation days with me, after our business and board meetings. Most of the board members attended the board meeting, which turned out to be heated and contentious; and as it progressed, I decided, in a fit of fury, when it was obviously the board wanted to follow the "Kendall" philosophy, to resign. I left the board meeting and returned to the hotel We left the Hilton, where we were staying, and flew back to D. C. the next morning. Kendall was promptly promoted to the Presidency of the Company, and begun to reshape the Company, initially attempting to move the Company toward selling educational and financial packages, which was much of his background expertise, he experienced while working in the Midwest with Booze Allen as a consultant. Shortly, after I left, two of the early key people I had hired also resigned from the Company, Jim Collins and Walt Scott', both whom I had hired and who had joined before the arrival of Kendall. It is interesting to note how history evolves and events change over the years. Kendall did a remarkable job in building the Company over later years and successfully selling the Company to Affiliated years later, building on what I started. But truthfully, he was never a co-founder of the company or one of 4 co-founders as is often quoted in the technology history books. The Company was designed by me and Jim Trawick originally, and when Jim dropped out, it was Easley, who replaced Trawick and then the two of us that co-founded CDSI, not Kendall. Kendall appeared on the scene, shortly later, after Jim Collins and Scott, but only when he was assured he would have a guaranteed salary. The entire public offering of the Company was started and completed by me through Charles Plohn, with Kendall in a peripheral way only preparing the financial data necessary for the IPO. If this section of my story sounds bitter, it is not meant to be; it is only mentioned to correct the history books! Kendall was brought in to handle the accounting matters after we started. The really significant, and most important individual in the beginning days of the Company, was Jim Collins, who was a gifted technical person, with a prolific ability to write cogent and powerful technical proposals to win contracts. Jim was the third employee hired by me in the Company, followed by Walt Scott and then Kendall. Collins and Scott had both worked for me, and later again, at C3, Inc. and Computer Equity Corporation, where they both would contribute heavily to the success of both those Companies. Interestingly, just before I resigned from CDSI in Paris, I had been talking with Jim Trawick, the original co-founder of CDSI, to join the Company, since it was now past the stage of a basic startup, and he was leaning toward that decision when I left. Jim did join the Company shortly thereafter, and was also a key person in continuing the growth of CDSI, as he also had a long successful background in writing, winning and managing contract work.

So, once more unemployed, in the summer of 1970, Pat and I returned to the States, and took a short vacation in Delaware to think through the next stage of our lives, and I am sure by now she had wished I had long ago elected to have joined the teaching profession. By now, we had four children, Glenn, Stephen, Christopher and Jacqueline, our beautiful, blond haired youngest child to support, so the vacation to Delaware had to be brief, so I could get back to earning a living. Since my oldest son, Glenn, had been a boy scout and wanted to camp out, my wife thought it would be kind of a good idea to take a tent and camp out at the beach versus renting a hotel room or two. I believe that her real intent was to camp out simply because we had little money and no job to return back too, so camping it was! We found a camping ground about two miles from the little beach town of Lewes, Delaware and pitched the tent, thanks to Glenn and his Boy Scout expertise and settled in for a couple of days, smack up surrounded closely by other fellow campers. Readers should understand that I am not the camping type;

so by the second night I was ready to pack up and leave; especially during dinner, when we were eating in our tent, and I asked for the salt shaker, only to have a strange hand reach into our tent with a shaker of salt, and a voice saying, "here Buddy, help yourself"! That kind of intimacy with strangers, I didn't think we needed, so the next morning we left the camp grounds, drove into Lewes; found a real estate agent and bought an old Victorian home on the Lewes Canal for $42,000 with not a dime in my pocket, financing it entirely with the owner, who threw in a house full of furniture for $500 more. I felt like my old friend and boss at CSC, Vince Grillo, and his tale of getting to New York and back without a penny in his pocket! That day I would have had trouble buying a shaker of salt, with what little money I had in my wallet. The kids were elated; a new "old house" built back in 1903, with tons of rooms, a creaky old elevator; a future summer home for the Ballengers for a lot of years into the future. Pat loved Lewes and ultimately wanted to retire there. A safe, quite, quaint town, discovered by the Dutch, and initially settled by them back in 1620, and later by the English, it was, and still today, an idyllic little fishing town, carbon copied from its English cousin, the town of Lewes in the southern countryside of England; a town we visited and toured several years later.

Once back home, I began to interview for a job, calling old connections, and hoping to land employment quickly. Unfortunately, I found out that being the ex-Chairman and President of CDSI, was a negative. Either I was considered overqualified for a position, based on previous titles, or I had the feeling during an interview with top management people that they were worried I could be hired and "steal" their job. I don’t know; I just needed work and couldn't connect. Meanwhile, Pat started back teaching elementary school, in of all places, a catholic school to help keep us afloat.

Chapter 12 – Informatics "A Year Lost"

At the end of my wits trying to find employment, I finally found an opportunity working for one of the largest software firms of the time called "Informatics", a software company that had built a software package called Mark IV, a general purpose business package designed for IBM hardware. They were beginning to expand into a myriad of software and hardware areas and thought I might be a perfect fit for their expansion needs. I first interviewed with the President of their Eastern Region, a friendly individual, Dick Lemons, who was intrigued with a new idea I had been thinking about for a while - Systems Integration, where we would design turnkey computer systems, marrying computers to "foreign" but less expensive peripheral devices, such as tape drives, discs, drums, cards, scanning, and others being produced by different manufactures, and if necessary, we would design and develop software drivers and mainframe software to interface equipment at less cost, thereby competing against large manufacturers, with carefully designed total systems, thereby meeting a customer's total requirement. Lemons was intrigued with this concept and arranged for me to meet with his bosses, Dr. Bauer, President and Werner Frank, the Executive V.P. of the parent company to discuss the concept. Bauer and Frank were an interesting duo to meet, quite different from the top officers at CSC. Bauer, a PHD mathematician, along with Werner Frank, also a mathematician had founded Informatics General in 1962 and ran the company until it was sold in 1985 with revenues of $200 million and approximately 2800 employees. An extremely detailed outside person, who enjoyed writing, speaking, and sitting on computer advisory boards. Bauer was less a hands on manager, leaving those duties to Werner Frank. Werner had immigrated from Germany when he was eight years old, grew up in California, and met Bauer, when they both worked for Ramo-Wooldridge a company that later merged with The Thompson Company to become TRW. They had started Informatics by becoming a subsidiary Company of Data Products, Inc., receiving a few stock options in Data Products, who financed the early beginning of the Company, with a $20,000 investment. It was not a normal way for entrepreneurs to start a Company, owning no stock in their own Company, but oddly enough, they both convinced Data Products to later spin the Company off in a public offering, returning Data Products $20 million for their original $20K investment, and Bauer and Frank received a minimum 5% stock ownership in the new Company.

When I first met the two of them and discussed my systems integration business plan, Frank was interested, but Bauer only vaguely so. At the time, Frank was the person researching and planning for new ways to grow the Company, especially in the hardware world, whereas Bauer was still "stuck" in the world of selling software packages and people support (low profit returns). In the end, I was reluctantly hired by Bauer, with the future promise of running a systems Integration company, but meanwhile brought on as a Vice President of Marketing to capture additional system analyst and programming "body shop" contracts. Informatics, at the time I joined, was still small, with revenues of about $11 million and profits of just over $1million. All management decisions were centralized and made by these two. It was an impossible career position for me; and after 9 months of working there, with no sign of the original promise to run a new division, I just walked out the front door and never looked back. Compared to the management and building team at CSC, Informatics' top guns were amateurs, and in my opinion, because of lack of foresight and sluggish growth, were acquired in a hostile takeover a few years later, by Sterling Software, one of Sam Wylie's companies. But once again, I was unemployed!

The only good thing that happened to me in my short tenure at Informatics, however, was I had become good friends with two of the financial people in the Rockville, Md. Office, a young guy, John Vazzana, 13 years my junior, and Frank Gaudette, who will enter my story later. John and I spent time together at lunches and after work would talk about starting a new Company together, and those conversations ranged for a while from taking on vending machine coin operated equipment to miniature golf courses, and the more crabs and beer we drank, the crazier the business ideas became. John and Frank stayed on at Informatics for a while, when I left, but John soon joined me in a new adventure that became one of New York's biggest financial hits in 1980, the year when Apple Computer and our Company, C3, became two of the four largest IPOs' along with Nike and Genetech that happened that year.

I almost forgot to mention that during my time at Informatics, my wife continued to teach at St. Camillus Catholic School in Silver Spring, Maryland, where, Glenn, Steve, and Chris, my three sons were now attending. True to her promise, she was raising the kids as Catholics, and going the extra mile by teaching there as well. I'm sure Father O'Sullivan had spy cameras hung throughout the ceilings of her classroom, ready to pounce, when he heard any criticism of church teachings, but to the best of my knowledge, he never reappeared! Our children were growing up fast and Jacqui, my daughter would be heading to grade school the following year, so we started to think about moving to a different area of Maryland, with a better school system for junior and senior high school students. So, during the time at Informatics, I bought two acres of land in Potomac and designed a new home for us, a beautiful little town, near the Potomac River, and once more started playing the role of an amateur architect. Money was still an issue for us, even though I now had a job and the only equity we had was the CDSI stock that Kendall and CDSI refused to let me register and sell; so the process and cost of my attorney was taking its toll on our lives and financial pocketbook. Ultimately, we were successful in getting my stock in CDSI registered, but not with the cooperation or help from the Kendall group. As I remember, the case that we presented to the SEC, and won, ended up becoming a significant case that loosened up the law relating to the 144 rule, that made it difficult to register unlettered stock, if the issuing company refused or "dragged their feet" in removing stock legends. All of this happening about the same time made it difficult to obtain a construction loan for starting the new house, and most banks turned me down, for lack of enough credit, and the fact that I had no prior building experience.

Luckily, I found a small bank, in Rockville, Md., with a sympathetic banker, who agreed to a construction loan, and began the process. As it turned out I couldn't find the personal time to oversee the project, so I hired a small builder and by the time I joined my next challenge, C3, the home was completed and our family moved to Potomac.

Chapter 13 – The C3 Story - A Seventeen Year Run

Almost as soon as I walked out the front door of Informatics, I had a call from an old friend at RCA, Dick Gasparie, who had worked with me years ago at UNIVAC, telling me of a little Company in Northern Virginia, C3, Inc., that was looking for someone to salvage the Company, since it had fallen on bad times, and the investors wanted help. Gasparie knew the Company and the people running it, since most had worked with him at RCA, and left to form C3 with financial backing from Homer Guldelski, a wealthy real estate developer. Guldelski had owned the land in Virginia that became Tysons Corner, and developed it with Ted Lerner, a local developer himself, and of late, the current owner of the Washington National's baseball team. The founding team that had left RCA to start C3, were led by "Bud" Shuster, Dusty Rhodes, and his technical team of David Sullivan and Arnie Shore, and Chuck Perko, Vice President of Manufacturing and field service, all still working for the Company, except for Dusty Rhodes, the President, and Bud Shuster, the Chairman. Rhodes had disappeared, moving to Florida and Shuster, who was from Pennsylvania, resigned from the Company and ran successfully for Congress, the next year, serving a number of terms in the U.S. Congress.

Back in 1968, when the Company started, there were several other companies designing and manufacturing key to tape and key to disc conversion devices replacing the old fashion key punch machines that created punch cards as the input medium used to transfer data to computers. These new devices now allowed data to be fed off line thru the use of magnetic tape or directly to the computer via "key to drum" devices. So C3 was born, designing these devices, around the new Data General mini-computers, competing against more recognized names in this industry, such as Mohawk Data Systems, the largest and most widely known, at the time. Little did I know about the sad shape the Company was in when I agreed to join, and I always use to tell investors, if I had known, I would have started a new C4, as opposed to inheriting C3, with a negative net worth of over $3 million when I first showed up for work.

At any rate, Gaspari introduced me to one of the senior Directors of the Company, Marty Seldeen, a cousin of Homer Guldelski, and also a well-known local developer, who built Wheaton Shopping Center, one of the first major mall complexes in the Washington area, as well as major housing developments throughout various Maryland Counties. Marty liked me from the start, although at this point in the Company's downward spiral, he would have been happy to have anyone with computer knowledge take over control of the Company. As happened often, in those days, a number of investors placed money in computer ventures and were then held hostage by the technical types, who “blew” thru investor money, knowing full well that investors would not know who to turn to if things turned sour. And this is what had happened to C 3! Seldeen and Guldelsky had invested over $3 million in the venture and most of the money was gone. They had promoted David Sullivan the Presidency, replacing Rhodes, who soon developed a hostile, in your face, attitude with the Guldelski group and the company was heading toward bankruptcy. Why I joined this Company, I'll never know! I guess the challenge, and a chance to run something once again overcame good judgment, and after meeting Guldelski, I felt if I could orchestrate a turnaround, he might be inclined to provide us with more needed capital. The Guldelsky offer was not a flattering offer, but one I accepted. A starting salary of $35,000 and a 15% stock option in the Company spread over 3 years.

Guldelski was a well-known real estate Developer and land owner throughout the Maryland, Virginia and D.C. Area. Also, he was the owner of Contee Sand and Gravel, a huge excavation and cement operation servicing the local needs of builders and road construction companies. When I first met him, the general consensus of his net worth approximated $500 million, but you would never have known it, based on his style of living. He lived in a small 1950s' brick home in the heart of Silver Spring, Md., and I can never forget the first time visiting him. In his small living room sat a piano resting on an old oriental rug, with a major worn out area where the piano chair sat. Homer practiced frugality to the "tenth" degree, and expected workers and relatives to do the same. He was one of the icons in the Washington Community and a close business partner with others like, Ted Lerner, who I mentioned was his partner in developing Tyson's Corner in Virginia, one of the largest mega-malls in the country. He also was involved with Abe Pollen, also a developer, who built the Capital Center, and owned several major sports franchises, including the Washington Caps hockey team and the Wizards professional basketball team. The Guldelskys' were also known for their major contributions to local charities, and gifts to various schools, including the University of Maryland. His office, at Contee Sand & Gravel, was so small, that no more than three could fit within it at one time and he co- shared this office with his key employee, Charlie Haughton, a tough individual I had to deal with for years to come. One can only imagine how he felt, investing $3 million in C 3, a technology business he never really understood. He had invested only because his cousin, Marty Seldeen convinced him of the potential financial opportunity it offered.

Seldeen then introduced me to Sullivan, informing him that I would be taking over the Company with Sullivan returning to run the technical part of the business. It was not a happy meeting, but he reluctantly agree to do so, but discretely fought me for the next few months before I forced him to leave the Company. Speaking about leaving the Company, on my first day at work, I met with our lone accounting clerk and discovered that we only had enough cash to cover the payroll for two more periods, and at the time there were 18 employees and several leased company cars, plus all other overhead that had to be covered. The remainder of that day, into the night, was spent interviewing each employee, as to their importance; and by the end of the evening we were down to 8, including myself. It was not a pleasant start, and the remaining 7 people were not the happiest or securest group of people on the face of the earth. The Company's backlog of orders had dried up, and Sullivan had left in my desk, without telling me, a cancellation notice from the 3M Company, our only large customer, for the remaining key dozen disc systems they had previously ordered.

It was obvious from the first week at work, that it would take more than one miracle to save C 3. Bills were piling up, on top of those that had not been paid for over a year. The Company had attempted a public offering, two years back, and owed our attorneys at Weinberg & Green over $200,000 and the accounting firm, Grant Thornton, over $125,000 for work on the offering prospectus. Miscellaneous bills totaling several $100,000 coming due, and many overdue for months stacked up on my desk! Debts were something I decided I could worry about later. The first thing I had to do was find a way to speed up our revenue collection process and to convince our main customer, the 3M Company to renew their order with us, so we could meet payrolls and pay the rent. So we immediately implemented a 10% discount to any customer who would pay all outstanding bills within ten days, and many did, which kept the Company going for a while longer. The key, however, was 3M, and I came up with an idea that I thought might work. Jack Veale, my old boss at UNIVAC, was at that time, President of an optical scanning company, in Newtown, Pa., and who happened to have a distributorship relationship with the same Division of 3M that had just cancelled our contract. I quickly met with him; explained our problem; asked him to join our board with a "juicy" worthless stock option as compensation; talked him in to flying out to Minneapolis with me to meet with Scotty McArthur, the V.P. of 3M, who also was a friend of Veale's. McArthur was a Viking's football fan, so that weekend we flew to Minnesota; took him to the Sunday afternoon game, or I should say he took us, being a season ticket holder, and by the end of dinner, later that evening, Jack had convinced him to reconsider his cancellation with us. It was a gigantic favor that he did for us, or I should say for Jack. MacArthur had told us our product was not moving fast enough through his sales distribution group, and that they had over two dozen of our key to disc systems that were still in inventory. But because of his friendship for Veale, he agreed to take the final dozen and pay for them over the next 60 days. In addition to recapturing the 3M business, Jack Veale agreed to also sign a distributorship agreement with us, and although not wildly successful, Optical Scanning Corporation's sales force did sell a few over the next two years.

After about three months at the Company, I began to make some personnel changes in the Company. Dave Sullivan had left from the Company, leaving only Arnie Shore and Chuck Perko that were most familiar with our key-to-tape and disc systems. As Sullivan left, I moved Shore into his place, as the key technical support for software and Perko for hardware. Shore had been on the software development team from the start, so he was familiar with the system, but if he left, there would be only a couple of junior technical support talent. left; and he was loyal to Sullivan, not me. I had to move swiftly to close that support loop, and shortly did. I was able to locate and recruit a young man, Dr. John Korpi, with an outstanding software development background, who quickly absorbed the software workings of our equipment from Shore. Perko, was one of the remaining RCA people on our staff, repairing and maintaining our systems in the field. He was slightly older than the rest; a little quirky in his personality, but he did his job, and was not especially loyal, or in bed, with the old RCA crowd, so I learned to trust him over time. Along with Korpi, three other key players were hired, along with a secretary, Barbara Dempsey to handle the phones and type our proposals. Paul Kazek, a senior technical manager from CSC, who had also, became a close social friend of mine, from those days at CSC, joined as V.P. of Operations.; Leo O'Keefe, from UNIVAC, to write new proposals and lastly, Timmy Young, a neighbor of mine from Potomac as interim CFO. replacing the lone accountant, who left the company. Tim was a very conservative guy, who reluctantly joined, and was alarmed when he saw the books for the first time. He was bright, but a 9 to 5 pm kind of guy, who was really waiting for a government appointment to open up, that he had applied for; but he stayed for a while and for the first time since I arrived, we were able to organize a better financial system for the company.

"Clearing the Deck"

With fresh new personnel on board, I changed the direction of the Company from key-to-disc technology to the marketing of "systems integration" contracts, which I had earlier tried to convince Informatics to start. Speaking of Informatics, with Tim Young wanting to leave, I called John Vazzana, who had worked with me at Informatics, to join us as the new V.P. of Finance, and the day he did dramatically changed the fortune of the Company. I believe to this day that there was no better technology company team in the area than ours, at that point in time, and within another year we had cleared away our backlog of debt and begin winning a number of new contracts in the systems Integration business. The first thing we did, upon his arrival was to negotiate a settlement of the huge IPO debt with our lawyers and accountants. Barry Friedman, a partner at

Grant Thornton, and in charge of our account, was overly gracious in negotiating a settlement of the $125,000 we owed his firm, and after a few days of talks, he agreed to cancel the debt if we would provide him with 6 months of systems programming support on a contract that Grant Thornton had with another client. By then, Dr. Korpi had taken over the key-to-disc program support, and our plans were to release Arnie Shore to reduce our overhead, but with the Grant Thornton settlement, Shore was assigned for six months to Grant Thornton. Shore was never happy working for the new C3 team, and ended up resigning and going to work for Grant Thornton, a few months after this reassignment. Barry, never pressured us for further support to finish the assignment and I can only guess that he felt guilty about "pirating" one of our "key" employees and forgave the debt. For us it solved two issues; a major debt and ridding us of a less than loyal employee. Next John and I travelled to Baltimore and met with Richard Himmelfarb, the attorney that represented the Weinberg & Green debt of over $200,000. From the start, there was bad chemistry between Himmelfarb and myself. He was difficult to negotiate with, and if it had not been for Vazzana, that debt would probably still be owed. John stepped between us and convinced him that the Company was strapped for cash, but on our way back, and convinced him that rather than cash we should give him a block of stock that would be worth more in the long run, than trying to extract a reduced cash settlement from us. So, by the end of the day, we shook hands, and walked away with a cancelation of $200,000 more of our debt, by giving up 25,000 shares of C3 stock, which at the time, was really "monopoly" paper. It turned out well for the law firm, however, as years later, when C 3 went public that "paper" was worth over $750,000! We did have another "ace in our pocket" during the settlement meeting, and that was "Guldelsky"! Weinberg & Green's major client was Guldelsky, and Himmelfarb did not want to alienate Guldelsky, since Homer still owned C 3!.

John then took charge of negotiating our way out of a significant part of our remaining debt, reducing it by sprinkling a little stock; discounting our debt; stretching out payments and in some cases canceling debt completely, with our "paper" stock. John was amazing to watch in action. He had a sharp mind for numbers and tax issues, and was forceful, but pleasant in dealing with our debtors. He was good at convincing those that didn't believe we would survive to take 10 cents on a dollar, while convincing those that thought we might "make it" to take our "paper" stock in trade. So by the end of his first year with the Company, we had "cleared the deck" with a clean balance sheet, almost debt free! In between negotiating our debts, he also took charge of developing creative price proposals for bid submissions; and with Leo O'Keefe writing brilliant technical proposals it enabled us to shortly win a number of contracts with the U.S. Army.

While he was working on financial matters, I was busy trying to work out marketing relationships with various mini-computer manufactures for our planned entry into the systems integration business. I had several meetings with Ed DeCastro, the President of Data General, who was then supplying us with his "Nova" mini-computer, the core unit for our data entry equipment. Since we were a small part of his overall business, he was reluctant to offer any changes to our existing OEM agreement, i.e., reducing prices, and extending payment terms, so it became obvious that I had to look elsewhere. In order to grow, with our tiny balance sheet at the time, I had to search out other mini-computer firms and peripheral manufacturers that would offer terms that allowed us better pricing and time payment privileges that would coincide with our customer invoice collections. The mini-computer industry was fast replacing main frame computers by the mid-70s' so there were a number of companies that were interested in replacing Data General as our OEM suppliers. I had hired Leo O'Keefe because of his technical writing skills and his ability to select companies that were technologically superior to others; to boot us quickly into the hardware integration business. Leo had determined that two companies, that would fit our needs, besides Data-General, were Digital Equipment Corp. (DEC), a popular system, and a small firm by the name of Interdata, that was building a line of mini-computers, and looking for OEM customers to sell into the U.S. Government. John and I flew to Boston and met with Ken Olsson, attempting to convince him we could become a major customer selling DEC systems to the U.S., Government. Olsson was friendly, but they were becoming the number one mini vendor, and he wasn't about to offer a tiny company better terms than larger clients, so we parted and left for Logan Airport. At the airport, having a drink before leaving for home, I said "John, I was waiting for you to pipe up and say, Ken, if you give us a good deal, we'll give you a bunch of C 3 paper stock"! Knowing Vazzana, it probably would have worked. John was just so damned persuasive in convincing people to take our worthless stock, that I was beginning to think that's all we had to do was print more stock certificates. That philosophy changed quickly over the coming months however as the Company turned profitable, and we stopped giving stock away. As a matter of fact, during the next few months, I started contacting the smaller early investors in C3, other than Guldelsky, offering them $.35 a share for their holdings. Many were ex RCA employees, like Gaspari and Congressman Shuster; several lawyers, Dusty Rhodes, the former C3 President and most of them took my offer. Vazzana and I would share 50-50 on all these shares purchased, so little by little, we became legitimate stockholders in the Company. In addition, I made sure that all of our new employees got stock options in the Company, and I'm proud of the fact that over the years, more than 30 people that worked for me at CDSI, C3, and lastly Computer Equity Corporation, became millionaires, or multi- millionaires. In addition, at C-3, over the years, John and I created over 1100 jobs for people, which I believe supports the view of many economists, business leaders, and a number of politicians today, that small businesses are the creators of jobs- not the Federal Government!

Meanwhile, we were still manufacturing a few of the key-to-disc systems for the 3M Company and Veale's Company, in our tiny warehouse space in the same building where we had our Corporate Office. Our office was in a small office building in Merrifield, Virginia, with a mix of other small businesses. Down the hall from us worked an attractive young woman, who brought her little white French poodle to work, every day, and leave her office and return several times a day, Leo fell in love with her, and when he took a break from writing, he would often go down and visit with her. After a few months, we found out, she was a full time "hooker", plying her trade, via of her phone from her one room office. Anyway, while she was there, she kept Leo motivated, and his technical writing skills kept improving by the day! Barbara Dempsey, our Admin. Gal, an elegant and lovely woman just shook her head!

"A Major Win"

Finally, after months and months of fighting debtors, dissident employees, lawyers, making payrolls, and generally struggling to stay alive, our little ship began to turn away from the storm. Leo and John were able to put together an extremely competitive proposal to the U.S. Army for $1million worth of our key to disc, data entry systems. Several of the Army evaluation group were people that I knew from my days at UNIVAC and CSC, working with them for many years, and it helped us get a fair hearing and evaluation. The Army was always concerned about selecting small companies, especially ones' they did not know, so when we were notified that we had been selected for the award, there was one problem; because of our weak financial position, they gave us, a conditional award based on our obtaining a bank loan commitment to finance the $1 million contract due within FIVE working days!

Elated with obtaining the win, I called two of our Directors, Seldeen and Veale, asking for advice and help with obtaining a financial commitment letter fast! Five days is not a helluva lot of time to find that kind of money. While they started searching for a bank commitment, John and I met quickly with our small Virginia bank, and my banker, State National, that had helped me with my home construction loan, a while back. Neither would approve a bank line of credit for a $1 million contract, because of our size, so I pleaded with Seldeen to convince Guldelsky, who still owned most of the Company to convince his banker to commit. Veale could not help, so during those frantic hours, I called, Bob Hanley, at Federal Data, who now was successfully leasing computer systems, and had good banking connections, offering him a percentage of our contract profits. It was an insulting meeting! Instead of agreeing to help, or just simply say no, he sensed we were in a desperate state, and offered to buy the entire Company for $85,000. Leaving his office, I called Seldeen and emphatically told him, it was on Guldelski's shoulders. Either he financed the Company, or we would lose the contract. A quick meeting was arranged by Seldeen and Guldelski's, Administrator, Charlie Haugh, in Baltimore, with Guldelski's lawyers, but not with the Bankers. Not his Bankers! I couldn't believe it- we were down to a couple of days from having to notify the Army, and we are spending almost an entire day with attorneys!

Further, I then found out that Guldelski had put his nephew, Irwin Guldelski, who worked for him, in charge of C 3 affairs. Irwin had no interest in C 3, or actually any of Guldelski's business interests. He was simply on the payroll, and did virtually nothing, except play "tennis" almost full time. Complicating the issue, was that Himmelfarb, the attorney, was in charge of deciding, whether Guldelski committed, or not. John and I waited in their lobby, most of the day, until they both walked out of their conference room to inform us that Guldelsky would not commit to the guarantee! We were flabbergasted, and we both exploded in anger at both Irwin and Himmelfarb; saying we were resigning unless they agreed to sell the Company to us immediately. Irwin said he would talk to Homer, that evening, and Vazzana and I left and headed home. On the way home, furious at the waste of a precious day, and thinking through other alternatives, John suggested we give Bill Thompson, who worked for Hanley, a call. Bill was one of the senior management people at Federal Data, who handled banking relationships for Federal's Government contract needs, and had sat in on our original meeting with Hanley. We met with Thompson, the next morning, in Bethesda and after an hour or two, discussing the details, he agreed to introduce us to one of their Bankers, at the Bank of Bethesda, who was familiar with, Government contracts and had financed previous ones. There was a cost, however, I had to ultimately give Thompson 40,000 shares of my own personal stock, if we bought the Company from Guldelski, and also a Board seat which I agreed to, and so we got the Bank letter commitment, and that's all that counted. That stock turned out to be worth over $5 million, after stock splits, the day we went public, and Thompson "piggy-backed" all his stock that day of December 17, 1980, one week to the day, after Apple Computer went public. As hefty a cost that was; at the time we had no other options, but the cost seems minimal today, when, with Thompson's help, we further convinced the Bank, to loan John and myself $75,000 to buyout Gudelksi.

The next day, we met with Himmelfarb and Irwin by phone, and told them 75K or else we walked. Guldelsky, by now was tired of the C3 issues and agreed to sell, keeping a preferred position in the Company, worth about 15%. John and I split the rest; with me retaining a larger percentage than John, because of the Thompson commitment I had made, and the added earlier time I had been at the

Company. Next stop was the Army, delivering the Bank commitment, which started us on the road to becoming the first, and one of the largest Government contractors in the systems integration business over the next 15 years! A Company that I joined in 1970, with a negative net worth of over $3 million and 8 people, growing to $100 million in revenues and over a thousand employees, when it was sold in a leveraged buyout in the late 80s! Oh, I almost forgot to mention, in my last meeting with Himmelfarb, my words were more than harsh, and later came back to haunt me. Years later, when searching for an investment banking house to take Computer Equity public, we ended up making a presentation to Legg Mason & Company, in Baltimore, Md. And as I entered the room, there was Richard Himmelfarb, now Executive President of Legg Mason, and the key guy at the firm for IPO approvals, smiling at me like Hannibal Lecter, in "Silence of the Lambs". Needless to say, we continued our IPO search elsewhere!

"The Integration Years"

As it turned out, the Data Entry Army win, was the end of our emphasis on marketing key- to-disc equipment. We now turned our attention to the mini-computer marketplace, which had begun in the early 70's, and to realign our staff to compete in that world. Paul Kazek, who I had met and befriended at CSC, and was now our V.P. of Technical Operations, had become a disappointment. Paul had worked to get a real estate license, when he first started to work at C3, as a backup in case C 3 never worked out, and he spent most of his time, or our company time, selling real estate, instead of C3. He never really fit into our hardware world and shortly later, John and I decided he should go; so with pressure from both, he resigned! Instead of replacing Kazek, with another V.P. Of Operations, I felt that we needed additional senior technical help, with writing skills to assist Leo O'Keefe in turning out more proposal bids, as opposed to another expensive manager. Well, the recruiting skies opened up, and we found another "star", Kathleen McWilliams, who was a partner with her lover, in a small local computer company; had just broken up with him, and was looking for a new opportunity. Kathleen was brilliant. She was a classical pianist; studied classical piano at the University of Heidelberg, in Germany; spoke fluent German; skilled as a programmer and writer in a host of different computers, and lastly, but not least she loved working for small companies! So, she joined us in 1973, and for the first time in the Company's history, we had two piano players working for us; Leo, with his "Mel Torme" voice and Kathleen, who could match him with her beautiful classical music renditions.

I had mentioned in an earlier chapter that John and I had begun calling on various mini-computer companies, searching for those that could optimally help us win Government and Commercial procurements. I also had mentioned that one of these companies was Interdata that built a new line of 16 bit computers competing against DEC. the giant in the industry at the time. And luck would intervene one more time for us; a friend I worked with, now a long time ago, by the name of Gene McLaughlin, was the Washington regional manager for Interdata. I must pause for a second here, and comment on the number of "friends" I keep referring to, throughout this story, that helped us. It probably sounds to you like John and I were the "Dale Carnegies" of the technology world with all of these friends, but there truly were many that helped us, and along with a talented staff of people, contributed immensely to our success.

Anyway, McLaughlin negotiated an OEM agreement with us in the mid-70s and we began to compete using Interdata computers, connected with a variety of computer peripherals we purchased from other vendors, that were compatible with Interdata, or if not, we set about designing the software and hardware to make them so. Not only, by then, were we a systems integrator, but also now we were a systems engineering firm, designing custom software and hardware. More talent began to join us; a creative design engineer by the name of Brice Eldridge; Ed Spear, an ex Interdata employee, who would become our Vice President of Operations in Europe in a short period of time, and others, like Steve Wessel, also from Interdata, our future Vice President of Field Operations and still with us from the beginning, Chuck Perko, now Vice President of Manufacturing. We began winning a number of procurements, quickly expanding our operations, and moving to Reston, Virginia in the mid-seventies into larger facilities. Two major contracts during this timeframe occurred, that skyrocketed the growth of the Company. A multi-million dollar contract with the American European Independent

School System, headquartered in Darmstadt, Germany, followed by a huge $25 million dollar contract with the U.S. Navy that grew over the years to over 100 systems at Navy Bases around the world. As we expanded, the Company's reputation did as well, and during the seventies, we were selected by Inc Magazine twice, one year as number 256 of the top 500 small business corporations in America, and once later as number 92 of the top 100 small companies in America, lists difficult to make, and we did it twice!

I should mention here, that as our reputation spread, EDS, a company started by Ross Perot, which initially started in Texas as a firm processing medical claims and billing for insurance companies, as well as State and Local Government contracts, contacted us, and shortly thereafter, Ross Perot himself, appeared on our doorstep and offered us $3 million for C 3! Wow, in four years the number had gone up from the $85,000 Federal Data had offered us to $3 million from Perot. We thanked him politely for considering our Company; turned it down, and continued growing as a private entity. Only one other time were we approached by an acquirer, and that was in late 1980, while we were preparing our public offering with DLJ and E.F. Hutton, TRW talked with us and wanted to acquire us for $65 million, if we would abort the IPO. We did seriously consider this offer. John wanted me to drop our IPO, arguing that selling to TRW would be instant guaranteed wealth, versus the uncertainty and longer period of time to accrue it under an IPO. My feeling was that the public offering, when it occurred, would be well received, and in the long term, we would do better. But more on that subject, later! We were still in 1975 and ready to implement our European School System contract.

When we won the School contract in 1974 for a multiple number of computer systems throughout Western Europe, we needed someone to manage the operation overseas that knew the Interdata systems, and Gene McLaughlin, suggested a young guy then working for his Company, by the name of Ed Spear who would be the perfect fit. Ed's background was in the field engineering part of Interdata and he had an intimate "hands on" background with the hardware and software aspects of what we would be installing in Europe. Ed was married at the time, but shortly after, divorced, and with no children, was the perfect pick to take charge of our overseas activities, and that he did! We both packed our bags, flew to

Frankfurt to meet our new customer and C3's overseas operations began. Our systems in Europe were to be installed at each of the American High Schools for the children of American parents stationed at various U.S. Army, Navy, and Air Force Bases spread throughout England, Germany, France, Spain and England. Stepping back a second in time, part of winning the contract included our convincing the School evaluation committee that we could install and maintain these systems. Luckily, Mort Wimpie, one of the people I had met on my first trade mission, while at CDSI, had stayed in touch, over the years, and he was at the time headquartered in London, representing a New York IBM leasing Company. Mort was using a 3rd party British company, Systems Maintenance Service, Ltd., to maintain their systems, and he introduced me to them. After convincing us they could install and maintain our installations, we made them part of our proposal, that seemed to satisfy the Evaluation committee. So our overseas operation, started off with one employee, and a "Brit" contractor whom Ed had to personally teach, since most of their background was IBM.

Figure 17: Ed Spear at C3

I stayed a few days with Ed on that first trip, to meet the Director of the Program, Sam Calvin, who was an American citizen, married to a local German woman, Hilda. Sam was what we call back in the States - a “Euro-American”, a person who was happiest living in his case, Germany, and had no interest in going back home. He was a Government educator, assigned to running this computer program, teaching High School students how to program and use computer systems, but he had absolutely no computer background, whatsoever. Computers were still, new, relatively speaking, to the average person, and Sam was no exception. He was pleasant, but could be a difficult person to deal with, and as I left for home I knew Ed would earn his pay! Ed was perfect for this position. He was a take charge guy, a one man show with virtually no support behind him in Europe, except for the SMS Company, who as we were to find out, did not know a heck of a lot more than Sam Calvin about these new computers. We were given office space at Darmstadt Army Base, a small operation about 60 miles from Frankfurt, where Calvin was based, and Ed moved in, took over and stayed for the next two years, managing one of the largest complexes of Interdata systems in Western Europe.

Back home, C 3 business was exploding! It seems like the whole world wanted to own mini-computers. We were winning contract after contract, during the years following 1976. The School contract was followed by a major win for over $25 million with the U.S. Navy placing Interdata integrated systems at Navy bases worldwide. And as we grew business, so did our worldwide people organization grow to almost 800 by the time 1980 approached; with a full time training center, for customers and employees, as well as a 40,000 square foot manufacturing center in Reston, Virginia. In addition, we had built a solid engineering and design center under Brice Eldridge management that created a multitude of devices that were compatible with mini-computers we were selling, and in several cases actually reverse engineered computers themselves. Our field maintenance and service organization had grown to several hundred employees worldwide, so that credibility and size were no longer an obstacle to winning large scale programs. C-3 had a wonderful competitive advantage over the main frame and mini-computer manufacturers by the fact that when the Government released a bid specification we could be technically one hundred percent compliant by designing a turnkey system, providing the client with exactly what he wanted, whereas a manufacturer like IBM, or Digital Equipment Corporations submitted their "brand name" products, which often times were either an "overkill" or in other instances an "under kill" bid proposal, causing them to be higher priced on overkill projects, or technically disqualified on under kill computer systems.

Our Company was in what I would call the "sweet spot" of the industry. Profits were soaring into the millions of dollars annually, and suddenly investment bankers were beginning to take notice. In 1977, I received a call from a gentlemen by the name of George Ferris, who was the President of Ferris Baker, Watts, a regional brokerage and investment company in Washington, D.C. George had been following our Company, and our successes, and wanted to discuss doing a Regional public offering of our stock. After several meetings and much conversation, it was decided that since Ferris mostly did small IPOs' or was an investment house who participated in larger syndications, handled by larger institutions, that at the right time, he would introduce us to the "right" people in New York, which he did in the later part of 1977.

The pace of our growth was staggering from the mid - seventies on; John Vazzana was promoted to Executive Vice President and became a Member of the Board of Directors. A new subsidiary company, Tempest Technologies, Inc. was formed, and placed in the hands of Dulaney Blaine, a seasoned technology and financial veteran that we recruited. Tempest Technologies was created to take advantage of the fast growing Government computer security world. In classified agencies, like the CIA, NSA, or the Defense Communications Agency (DCA), the government was concerned with classified information being captured from unprotected computer systems. As a result, they initially built steel encased shielded rooms to place classified computers in. Although this solved the problem, the secured rooms were expensive to build, heating problems from the confined closed rooms became problems, and the cloistered working conditions begged for a better solution. Tempest Technologies offered that! We were fortunate in stumbling across several electrical design engineers that worked for NSA that had created secure technology that could be placed within our computer systems and peripheral equipment that would block any incoming spy radiation beam or detection rays directed at the equipment. As a result equipment could be place in a normal computer room.

Over the next two years, additional significant wins were awarded by the U. S. Coast Guard for an initial $35 million, and the U.S. Army Supply Command (DARCOM), for over $65million. We had continued our relationship with the Interdata Company, marketing each and every one of their new product lines and upgrades. By 1977, most mini computer manufacturers were moving from 16 bit computers to 32 bit, and even a few, like DEC had announced a line of 64 bit processors. Initial contract awards to our overseas clients and the U.S. Navy were Interdata 7/16 bit systems, but as the months slipped by, we moved to selling more and more of the Interdata 7/32 bit equipment. By the end of the 70s', we were the largest customer of the Interdata Company, with a technical and maintenance staff that closely equaled theirs. The DARCOM contract was one of the last major contract wins with Interdata equipment, as new products, using higher speed and lower cost microprocessors were being announced. So by 1978, we began offering microprocessor computers workstations manufactured by Convergent Technologies, Inc., a Santa Clara Company. This initial system was called an IWS (Integrated Work Station), and was a creation by Allan Michaels, President and the Chief Technology Officer for the IWS. The IWS development was one of the early entries into the next round of computer evolvement, "Desk Top Computing", where Customers had their own desk top computers, tied to centrally located disc and printers. These systems, fore runners of PCs', became wildly popular; enabled us to compete and win our Coast Guard contract; and took Alan Michael's company from $0 to $400 million in sales over the next 4 years! Alan later joined the Board of C3, after our public offering in 1980.

Investment Bankers again began calling on us, urging us to consider a public offering - companies like Drexel Burnham, Hambrecht &Quest, Alex Brown, and thru an introduction by George Ferris, the firm of Donaldson Lufkin & Jenrette. Again, we decided to hold off going public until late 1980, but established a line of communication with each, and a few others, during that period, updating each with ongoing company progress. Everything right was happening to us! Nothing could go wrong!

"A Cry from Home"

But it did! For the last few chapters I have been focusing on the Company, our people, and the Industry, but neglecting to reflect on what was happening in my family life. Earlier I had mentioned that Pat and I had moved to Potomac, Md., a lovely little village. In the Northwest part of Maryland, not far from Washington, D.C. The four kids were growing up; the oldest two. Glenn was away at Tulane University and Steve about to begin his first year at Newbury College in Columbia South Carolina. Christopher and Jacqui were now going to Walt Whitman High School in Bethesda, Md. Pat had stopped her teaching career, once again, when we moved to Potomac, to manage the family; and as anyone with three or more teenagers in a household can tell you, that is one of the most courageous, challenging, and difficult professions on the face of the earth! Again, like with past companies, I had to spend an inordinate amount of time, away from home, traveling around the Country meeting with customers, and employees, as we continued to expand our operations.

I happened to be in Austin, Texas in the middle of the summer of 1977, visiting one of our Interdata customers at the University of Texas, who was designing a new general purpose interface device for us under a sub-contract, when I had a call from one of my neighbors, who had dialed 911 to have, my wife ambulanced to our local hospital. Pat had suffered a massive intestinal rupture and had to be operated on immediately. I panicked, and caught the next plane back home, and went directly to the hospital, where she had surgery, and was recovering. Neither of us had ever suffered any major health problems before; she was still young and vibrant, full of life; and it just seemed unreal that this had happened to her. The kids were terrified, as was I, and for the next several weeks, I took off from work to be with her and the children, when she came home from the hospital. From the day of that first 911 call, our family lives changed forever; and over the next two years, she would endure three more major surgical operations, and a long separation from her family and home, before she died on March 17, 1979. Those two years between 1977 and 1979, were two years of living “hell" for her, and a horrific, stressful and agonizing time for the children and me. Unfortunately, the rupture break turned out to be malignant, and slowly spread throughout her body. The two of us were desperate to find a cure for her disease and searched and talked with a number of Doctors and Research Scientists at NIH, and elsewhere, trying to find a treatment program for her that would slow, or miraculously cure her cancer. We were told that that there was a number of experimental programs underway at MD Anderson, in Houston, Texas, and Senator Ted Kennedy was promoting more funding for cancer research projects, both there, and elsewhere around the country, and had become a strong advocate of providing cancer patients with more research and treatment programs; tackling head on, the search for cancer cures. In the Congress, Congressman

Edwards from Georgia was supporting the return of a lone scientist, Dr. Laurence Burton, who had left a promising oncology immune-augmentative program at St Vincent's Hospital in New York City, and moved it offshore to Freeport in the Bahamas. He was under pressure from the American Cancer Society and the Food & Drug Administration; because he was treating patients outside the rules and regulations of the FDA. Burton had some early successes in curing or dramatically slowing different types of cancers and after talking with staff members of Congressman Edwards, and Kennedy, we decided to fly first to M.D. Anderson Hospital, in Texas and then to visit the Immunological Center in Freeport.

Before we left for Texas, I put John Vazzana in charge of the Company, while I was gone, and left my three children, still living at home, with Delia, our housekeeper from Bolivia, to look after them. Shortly thereafter, we arrived in Texas for our appointments at the Hospital, spending two days with different research groups, before being told that there was no "miracle" treatment therapy that could slow or cure her disease. The night before we left Texas, Pat and I talked about whether we should go to Freeport or not, and without hesitation, she insisted on going; but first stopping off to see Glenn, who was at Tulane University in New Orleans. She was elated to see Glenn, and the old Irish sparkle in her eyes came back for that short weekend we spent with him, before leaving for the Island. Freeport was a pretty little Bahamian town, just east, off the coast of Ft. Lauderdale, more a resort haven for vacationers and gamblers, rather than a cancer clinic for patients. It is hard to describe the fear and uncertainty we felt; as we drove around the Island after checking into our hotel, because both of us knew that the treatment program would last, close to a year, and understood that patients had to live on the Island for the daily Injections that had to be self-administered. That meant a separation from her family for a long period of time! The daily injection serum given consisted of four proteins that had been isolated from human blood, and based on a patient's daily blood analysis, reconstituted in varied combinations, so it would be impossible to enter the program and be treated at home. It had to be done there! Burton felt that, used in the right combination, these proteins could restore the normal immune function in cancer patients. I wanted to go home; it was heartless to think of leaving her on this Island for almost a year, being treated with a controversial immune treatment process. There was further concern because there was only one other Doctor on the Island, if she needed emergency medical help, and one "native" hospital, that he worked out of; eerily, named the "James Rand" Memorial. How odd that was! It turned out that Rand, the President of Remington Rand UNIVAC, vacationed here, and had donated the Hospital, now run down, and poorly staffed by locals. Pat, however, wanted to stay. This was the last hope she had for a therapy that just could, maybe save her life! So Monday morning, we were introduced to Dr. Burton, and she entered the program!

The rest of the week, while she was being indoctrinated into the program, I went looking for a condominium to rent, and a car to lease for her to travel back and forth around the Island. Next we planned how to arrange for her to have family or friends, and myself visit enough days so that she was seldom to be left alone. By the end of the week, we had moved into a 3 bedroom condo on the water, plenty of room for friends, and rented a VW "Beetle" for transportation. I stayed with her, for another two weeks, before leaving to get back home to be with my kids and checking in with Vazzana and the Company. The tearful day I left, her best friend Jean Mitchell had arrived from the States to be with her, until I returned the next week with Jacqui and Chris, to visit. This started months of my sending family, friends, and myself commuting back and forth to Freeport. Normal family life insanely twisted and destroyed, in a few short months! I cannot describe, in this book, the rage and hopeless feelings that we had over her illness, and the tragedy that had struck our family. The night before I left, holding her in my arms, both of us sobbed uncontrollably, Jean Mitchell, a childhood friend of Pat's, was the perfect choice to be with her, as I left. Jean was compassionate, a terrific listener, and confidant that got her through those initial first weeks; but it was our kids she needed most, and over the coming months I did all I could to arrange school schedules, so they, one by one, could spend time with her.

Steve and Glenn were away at school, so only Chris and Jacqui were still at home, with Delia, our housekeeper, who was a blessing sent from Heaven! Delia had just turned thirty; was young and efficient, and amazingly in a short period of time had taken over the running of the Ballenger household. The kids loved her; she was like an older sister that spread love thru the household. For years after she left us, she stayed in touch with me, as well as the children. She was an immense help, allowing me to again focus on the Company, and the time to travel back and forth to Freeport, sometimes with family, or friends; sometimes by myself.

"IPO Courtship Continues"

Shortly after returning to work, George Ferris, invited me to join him for lunch at his office in town, and joining us was a Senior Vice President, from Donaldson, Lufkin, and Jenerette (DLJ), Sabin Streeter, who was in charge of uncovering new Companies, unique in their makeup and message, and a history of profits that he could attract to the public marketplace. He was intrigued with ours! He had never taken a Company public where the majority of its revenue came from State and Federal contracts, and was not familiar with the "integration" software and hardware strategies we used to capture awards from much larger and well-known Companies, such as DEC, Data General, and IBM. So, the IPO courtship with DLJ began, and picked up steam as we moved closer and closer to a public offering date. John and I had many discussions on how best to manage and grow the Company, and I finally agreed that he would continue to run day to day operations of the Company, and I would continue handling the investment banking and marketing piece of C3. John and I had a close personal and business relationship together; He was sensitive to the stress and pressures that our family now had inherited, and he did his best to help. As I mentioned in earlier passages of this book, there were several other Investment Banks that were interested in C3. Drexel Burnham, pursued us to the end, and we came close to dropping all others and placing ourselves in the lap of Drexel. It was a time when Michael Millikan's star, as the Junk Bond King, was at its peak with Drexel; and Stan Joseph, the President, doing deal after deal, using Milliken to help in closing them. Millikan was a brilliant guy, very personable; and even though his forte was the bond market, Joseph used him to help close equity deals; so Milliken met with John and me on two different occasions, once in New York, and once in Virginia. It was enthralling conversation to listen to him tell us how he could debt finance hundreds of millions of dollars for us, once public, to begin acquiring Companies, immediately, that overnight would explode our growth. He really wasn't interested in IPOs', he saw Companies through the eyes of "Junk Bonds"! Alex Brown, a Baltimore firm, expressed interest, as did Hambrecht & Quist, a Silicon Valley technology banker. Others like E.F. Hutton, and Merrill Lynch, a contact made through Bob Ryan, listened to our story.

End Of Freeport Commuting

Before Pat and I had left for Freeport, her Doctors, at Georgetown, had tried to discourage her from going, citing all the logical reasons for not going: lack of quality medical care, the FDA approval issues, few papers that Burton had published, relating to the treatment, and separation from her family and friends. Seeing that they couldn't dissuade her, they did one last cat scan, so they could compare the progress, if any, the treatment would have on her disease, when she returned. Over the months, she settled amongst other patients, and quickly made friends with a number of them. People from all over the world, had heard of Dr. Burton, and were flocking to the clinic, hopeful for a cure! Many were terminal cases; a number extremely Ill from the disease; and physically weak, but nowhere else to go. For many, it was too late, and a few died on the island, while being treated. After about four months in Freeport, Pat wanted to return home for a few weeks, so I flew to the Island, and brought her back! She was ecstatic, when she got back home; the kids were excited; even "Tippy" the dog sensed that this was a happy occasion. Friends and relatives visited often over the next two weeks, and for a few days, it seemed that our lives had somewhat gotten back to normal. When it came time to go back to Freeport, we were both hesitant as to what to do, so we decided to go back to Georgetown Hospital, to visit with the Doctors that had scanned her just before she had left for Freeport. Once there, they rescanned her, and the Doctors were astonished. Her scans showed signs of necrosis in the tumor, suggesting that it was dying. It was an unbelievable encouraging sign that Dr. Burton's treatment might possibly be working, and without a second thought, she opted to go back to the Island,; and two weeks to the day, we boarded a plane and headed back to Dr. Burton, hoping and praying that the treatment was working.

It was now early fall of 1978, and once again we began the ritual of friends and family traveling to Freeport, over the weeks, watching over her, offering encouragement, and co- mingling with new friends Pat had met on the Island. One such friend was a young priest, Father Guido that Pat and I met, when we attended mass at the only Catholic Church on the Island. The Church was in a black, impoverished section of Freeport, that contained both the Church and an elementary school, and Father Guido was the Assistant Principal of the School. Guido, who was from Italy, had been a priest assigned to a Parish in Rome; and after several years, when he began to doubt his commitment to the priesthood, he was told that it would be in his best interests for the Church to send him elsewhere for a while to "clear his head" and hopefully to reconsider his decision to leave the priesthood. So Father Guido was sent to Freeport, stripped of his priestly duties temporarily, and given the position at the School. He became a close friend to both Pat and I, and over the months, and he became a wonderful spiritual advisor and support arm for Pat. Later, when we had left the Island, he would visit us in Maryland from time to time, always with a bagful of delicious meats from Italy, that his mother would send him. Guido, a troubled, now defrocked priest, ended up moving to Kentucky, and in a last letter mailed to us wrote that he had married a former nun, who was teaching psychology at the University of Kentucky.

Thanksgiving was spent that year in Freeport, and since Freeport didn't celebrate Indians and Pilgrims, there were no turkeys raised on the Island. it was either pork or chicken, so the kids and I smuggled a turkey through customs, and we all celebrated the day with turkey, coconuts and papaya, and fried plantains!

Pat had begun to experience a good deal of pain, around this time, as well a loss of weight, so shortly after we decided to come home before Christmas, and to have her physically checked over at Georgetown. Tragically, the new scans showed that the disease was not dying, but growing steadily, and her pain becoming more pronounced than ever. Watching her get thru Christmas with us was cruel. She never complained, hid her pain in a heroic fashion, and always had that warm Irish smile for us, watching the kids open up their "Island" presents she brought back with her. Shortly after Christmas, she was operated on, once again, for pain relief; by then all knew that her time left with us was limited. Most of it in and out of Georgetown Hospital. The additional toll on our family life was heavy! All the joy, happiness and fun that had filled our lives for so many years vanished, replaced with a sadness and bitter, "Why Mom, Why us" permeated our conversations. Pat died, in Georgetown Hospital, at 7:30 in the morning, on March 17, 1979, St. Patrick's Day; a day God picked for a beautiful Irish Mom and beloved wife. She was gone!

Return to work

I soon returned to work after Pat's death, but reworked my schedule to be home in the early afternoons, when Chris and Jacqui would be home from school. Delia, our housekeeper, was still there, but I wanted to make sure the kids knew I was there for them as often as possible, friends continued to stay in touch for a while, but life being what it is, we all have our own needs and life to live; so shortly after, the phone stopped ringing, and letters stopped coming. It became a lonely, sad life without her. Thank God for the kids and work; I think both kept me from a physical breakdown. John Vazzana was a special friend, and support arm for me; not only did he do a marvelous job in running the Company during my many absences over the last two years, he was a special good friend, or maybe a special "good uncle" who cared about my kids. He and his wife, Cheryl took Jacqui under their wing, and did their best to soften her loss.

It was rough getting back to work and rebuilding a new life, but you deal with what God "deals" you and do your best to move on, in a positive way. It took me awhile to catch up with what was happening in the Company, and I found that John had done a magnificent job in maintaining our revenue and profit targets, and keeping employee morale high. Financial news outlets, i.e. newspapers, magazines such as Forbes, Inc 100; technology articles began to emerge country wide talking about C3 and their "integration" approach in the computer world, expanding our network of prospective customers; and access to talented recruits for our staff; as well as a flock of interested acquirers, including several of the mini-computer manufacturers. C3, in addition to my kids, became a large part of my family, and truly kept me from breaking up emotionally. There was always something new and exciting in our business world, which John and I had agreed to share in equally. I was always the major stockholder in the Company, based on my longer tenure there, as well as the one who “ponied" up the added shares to Bill Thompson for the earlier much needed bank connection. But in future deals, we shared equally; all stock we bought back from initial stockholders" we split equally. The same for a large block of stock that we repurchased from Paul Kazek in late 1979. Additionally we formed a new Company BALVA, that purchased U.S. Navy leased computer systems under contract with C3, providing C3 with instant earned income and revenue, while BALVA collected leased revenues and profits over the life of the leased equipment. "Life was good at the office"!

But it was still difficult and impossible to shake off Pat's death and the kids and I remained troubled with broken hearts, stumbling through school classes and work. They say children are more resilient than adults in handling tragedy, but I have seen firsthand how that is not the case for all; it was extremely difficult for mine. They had lost their mother; someone who was their "rock" to lean on; their pal that listened to their problems, the one person they could confide in, and the one they had genuine fun with! Fathers just cannot fill that void, it is a role that mother's own, and have exclusive rights too! Each of my children reacted somewhat different to their loss than the others. Stephen, of all the kids, was the most sensitive, quiet and moody of the lot. Without my knowledge, he had stopped going to classes his second Freshman semester at Newberry College and had taken a job as a rod carrier, toting bricks to construction sites, writing beautiful poetry and playing guitar in a band at night; before I found out and brought him back home. Chris, always quiet and reserved, kept his thoughts and feelings to himself; Jacqui, withdrew, and began to have difficulty at school, where counseling was not readily available. After meeting with her teachers at Whitman High, I decided to move her from Whitman, which was a large 1000 student body school, to a small private school in Potomac, that fall of 1979 where she could receive more in school counseling help. Glenn was still at Tulane but kept in touch with us all, and came home on school breaks. I know he missed his Mom no less than the others, but he always maintained an exterior upbeat personality, and covered his grief well.

Early spring arrived in D.C. and I was able to minimize my travel staying close to home, and my children. It was at this time I made perhaps one of the biggest and selfish mistakes of my life. Lonely and exhausted mentally from the two years of Pat's suffering and struggling to stay alive, my friend, Gene McLaughlin, concerned for me, suggested I begin to date again, and after a few persuasive conversations with him, he introduced me to a friend of his, Karen Kaub, who was an attorney working for the Rodinio Committee on Capitol Hill. She was an attractive woman, who had recently moved from Los Angeles to the D.C. Area; a graduate of UCLA law school, divorced, but no children, and obvious at the time, looking for a permanent relationship. I am not going to spend a lot of writing to discuss this period in my life, except to say, that I was lonely, mentally sick with sadness, and she just came into my life at a very vulnerable and opportunistic time. We ended up marrying that October, to the horror of the majority of my friends and family, and crushing my children. I don't think my children can and will ever forgive me for that mistake, because it further complicated and "screwed" up their lives. Steve began to drink extensively and dropped out of West Virginia University, after the Newberry College episode. I still hold myself accountable for his death in April, 1980, when he died in a motorcycle accident, two blocks from our home - only six months after my second marriage. Within a year, this marriage was over, but took almost another four years to legally end! In December of 1979, shortly before Steve died, another tragedy hit; my beloved father died of a massive stroke, so that in the period of just over one year, three of ours had left us.

Looking back, it seems remarkable to me that I could function at work, with all that had happen since Pat's death, but I pushed myself to and focused full time on the office. It really was the only way I could maintain a semblance of sanity in my life. The "buzz" of going public was loud and enthusiastic throughout the C3 offices by summer, when we announced that DLJ, along with E.F. Hutton agreed to take us public later in the year, and work began on developing our offering and preparing for the public road show, selling stock around the Globe. I had selected DLJ because they were known as one of the quality institutional banking houses in New York, meaning they had the connections to sell large blocks of stocks to insurance companies, pension funds and mutual funds. However, we also wanted to broaden the ownership of stock into smaller hands, as well, and so we selected E.F. Hutton, a recognized retail investment banker whose specialty was selling to individual smaller investors –to the guy or gal who would buy 100 shares. During this Sept.-Oct period, we also received our last acquisition offer before the public IPO from TRW, who offered us $85 million in stock and cash for the Company, if we would abort the public sale. The proposal was tempting, but we were so far along with DLJ that the uncertainty of added time and due diligence efforts were too great a chance to take; so that October, we issued a "red herring" indicating the Company would offer 1 million shares of stock for $18.00 a share, and began our pitch in various major cities in the U.S. and Europe. We split into two teams; John leading one, and I leading the other with support members from DLJ and Hutton joining each team. Over the next two months, we probably made a combined 50 presentations selling stock, with most of the stops being successful. A couple of interesting side stories I have to share with you on my travels! On a presentation to the IBM pension fund group, at one of the World Towers buildings in New York; the leading IBM'er in the group, walked up to me after my presentation; congratulated me and said he was extremely impressed with our Company, and intended to buy our stock instead of Apple Computer. He mentioned that Steve Jobs had made their presentation a week before us, and he liked our "pitch" better than Jobs. How about that! Both Companies followed each other on road shows and went public one week apart; Apple on Dec 10 and C3 on DEC. 17. As I had mentioned earlier, the four largest and most successful under writings of 1980 were Apple, C3, Genetech, and Nike Shoes. On another trip, when I got to Scotland, I met with the Director of the Scottish Widows and Orphans Pension fund in Edinburgh, at his favorite pub, where we had dinner and drank scotch together, late into the night. He, dressed in a kilt, long stockings, smoking a pipe, was more interested in hearing about my Black Angus farm, named "Brigadoon" in Virginia than buying stock. But by the end of the evening, he placed an order for a sizable block of stock, and I prayed that our stock would never collapse and be responsible for some poor orphan waif, or widow missing a meal or tossed out of a room because of our not being successful.

The C3 IPO

The grand finale of our road show days ended In London, during the first week in December, where John and I met up with the DLJ and E.F. Hutton staffs and celebrated the end with a large private party at the fashionable Simpson's five star Restaurant, near the Dorchester Hotel. The presentations had been highly successful, and the appetite for the stock had grown with each road stop, so that by the end, DLJ elected to move the offering price from $18 to $32 a share; almost double the initial commitment!

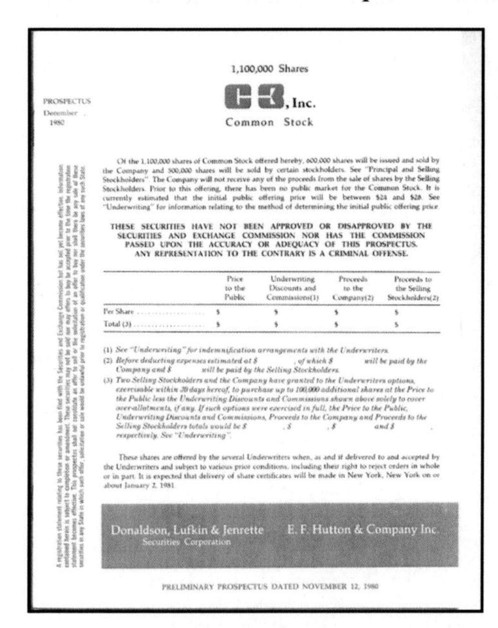

PROSPECTUS
December
1980

1,100,000 Shares

C3, Inc.

Common Stock

Of the 1,100,000 shares of Common Stock offered hereby, 600,000 shares will be issued and sold by the Company and 500,000 shares will be sold by certain stockholders. See "Principal and Selling Stockholders". The Company will not receive any of the proceeds from the sale of shares by the Selling Stockholders. Prior to this offering, there has been no public market for the Common Stock. It is currently estimated that the initial public offering price will be between $24 and $28. See "Underwriting" for information relating to the method of determining the initial public offering price.

THESE SECURITIES HAVE NOT BEEN APPROVED OR DISAPPROVED BY THE SECURITIES AND EXCHANGE COMMISSION NOR HAS THE COMMISSION PASSED UPON THE ACCURACY OR ADEQUACY OF THIS PROSPECTUS. ANY REPRESENTATION TO THE CONTRARY IS A CRIMINAL OFFENSE.

	Price to the Public	Underwriting Discounts and Commissions(1)	Proceeds to the Company(2)	Proceeds to the Selling Stockholders(2)
Per Share	$	$	$	$
Total (3)	$	$	$	$

(1) *See "Underwriting" for indemnification arrangements with the Underwriters.*

(2) *Before deducting expenses estimated at $, of which $ will be paid by the Company and $ will be paid by the Selling Stockholders.*

(3) *Two Selling Stockholders and the Company have granted to the Underwriters options, exercisable within 30 days hereof, to purchase up to 100,000 additional shares at the Price to the Public less the Underwriting Discounts and Commissions shown above solely to cover over-allotments, if any. If such options were exercised in full, the Price to the Public, Underwriting Discounts and Commissions, Proceeds to the Company and Proceeds to the Selling Stockholders totals would be $, $, $ and $, respectively. See "Underwriting".*

These shares are offered by the several Underwriters when, as and if delivered to and accepted by the Underwriters and subject to various prior conditions, including their right to reject orders in whole or in part. It is expected that delivery of share certificates will be made in New York, New York on or about January 2, 1981.

Donaldson, Lufkin & Jenrette
Securities Corporation

E. F. Hutton & Company Inc.

PRELIMINARY PROSPECTUS DATED NOVEMBER 12, 1980

Figure 18: C3 IPO Filing

The road trip had been exhausting for Vazzana and myself, but the end game reward was ours, and after ten years of hard work, timing and luck, the tiny, bankrupt C3 Company had risen from the business ashes with a recognized international presence. All the publicity was not 100% good. Two days before the public offering, Alan Abelson of Barron's, a popular columnist, for the paper, wrote a scathing negative article, inferring that our large profits from Government contracts were obscene and that the Government was paying for these highly inflated profits, most going to two individuals, Ballenger and Vazzana, providing undeserved wealth to them on the backs of ordinary taxpayers. Nothing could have been further from the truth, but Abelson, still writing for Barron's today, was and is known for his gifted sarcastic writing wit, and in the end his article on our high profit margins, I believe enticed additional investors to purchase our stock, not shy away from it.

Figure 19: C3 Board of Director: foreground John G. Ballenger, Chairman, left to right, George Kinsman, Jack L. Hancock, John D. Vazzana, Martin Seldeen, and Craig S. Sim.

On December 17, 1980, John and I were in the downtown offices of DLJ, when the market opened, and C3 started trading under the symbol CEE at $32, and ended the day over $40 a share. After a short celebration at their offices, Vazzana and I took a subway trip to Jersey City to pick up our $32 million check at the Jersey City DLJ office, returning on the same tram, with the check in our pocket. Remembering some of the "Occupy America" type faces on that train, if they knew what was in our pocket, we may have not made it back to New York! That night, we entertained some 200 guests, at The Tavern on the Green restaurant that included a mixture of family and friends, key employees, and directors, as well as all of the DLJ and Hutton IPO staff. Checks from the sale of "Green Shoe" stock sold by several of the Directors and key employees were given out at the party, including my $5 million "gift" to Bill Thompson for his introduction to the Banker back in Maryland, who actually saved the Company, in our darkest hour. Sabin Streeter of DLJ, our nervous, jumpy IPO Manager, who was hyper throughout the process, was fitted out with a straight jacket, we purchased at a local N.Y. Hospital, with the lettering "C3-32" emblazoned on the jacket, signifying the offering price. For DLJ that year, it was their most successful offering. Later they would produce a secondary offering of stock for C3 and also take our subsidiary company, Tempest Technologies, public. So the night of December 17, 1980, ended ten years of hard work, and luck, by a brilliant team of employees, who made it happen. The next ten years found different challenges and less luck heading our way.

C 3 Follow-up Years

As the year 1981 rolled in, we began to refocus more than ever on growing the Company, since the underwriting was no longer our primary focus. The IPO for all its value to the Company unfortunately brought us more at "integration" competitors. Our prospectus was sort of a "how to learning" manual for competition to copy our contract win methodology, and articles like Alan Abelson, all of a sudden put us under the watchful eyes of the Government. The era of the eighties became a period of time when the Government began to pursue major cases of fraud and waste in government contracting, brought on by the infamous $1500 "Toilet seat" story cost for military aircraft contracts. Unfortunately for a number of Companies, including IBM, as well as ourselves, it became a difficult and perilous time for companies conducting business with the U.S. Government. However, by the time I left the industry in 2002, there were probably over 200 firms, now directly or indirectly offering, and competing for Government and Commercial integration contracts. A new industry had been born!

I should stop here for a moment to highlight a luncheon that I had with Homer Guldelski, shortly after the Company had gone public. I had always gotten along well with Guldelski personally; it was his legal team that had made the decision for him not to further back C3, when we were desperately in need of bank financing, not Homer Guldelski himself. Guldelski ran a huge financial empire, personally worth over a $ billion by now, and C3 was just a small financial embarrassment to him, not a particularly "significant asset to provide him with 3 meals a day, and a roof over his head! So, Marty Seldeen, his cousin, who was one of our directors arranged this luncheon, at Guldelski's favorite luncheon spot, the Holiday Inn in downtown Silver Spring, Md. Guldelski was absolutely giddy with delight, when I handed him a check for a little over $3 million, for stock he sold in the offering. I'm positive; he never expected to see his investment again. He couldn't stop thanking me enough, and kept asking me what he could do for me. "Jack, what do you need?; a boat, a car, surely there must be something I can give you for doing such a great job"! When I politely declined, saying it wasn't necessary, he kept pressing, so. I finally mentioned that my daughter would be driving shortly, and she would be needing a car for school. "Well, let's get her a car! What kind of car would she like?" I answered back in a kidding way; "something safe and dependable, like a Mercedes diesel, maybe"! He answered back; "Nah, you don't want to get her that! Let's get her a Buick"! With that comment, He turns to Seldeen and says "Marty, get her a Buick", and with that, the luncheon broke up. Of course, the Buick never showed up but I always admired Guldelski; he was a self-made multi-millionaire, who donated millions to Charities, foundations, and the needy, but yet lived frugally his entire life. You see, he had always driven a Buick! Why would anyone need something better?

The year before we went public John Vazzana moved to Virginia, purchasing a beautiful horse farm in Middleburg, Va., and shortly thereafter I purchased a cattle farm, adjacent to John, that Senator John Warner had owned, when he was first married to Elizabeth Taylor. Warner was running for the U.S. Senate, at the time, and needed money to help finance his campaign, so he began selling off parcels of the estate, he had obtained from his earlier marriage and divorce to the daughter of Paul Mellon; so the piece that I purchased from him, I renamed "Brigadoon Farm", after one of my favorite musicals. In 1990, I built a lovely weekend stone home on the property, which has now become my full time residence and today, it is one of the town's most beautiful estates, dotted with ponds, and streams, wooded footpaths, wild turkeys, deer, zillions of red fox, overlooking the beautiful Blue Ridge mountains of Virginia.

In early 1980, I had decided to sell the home in Potomac that Pat and I had earlier built, and moved my family to another home in Potomac close by. Glenn, my oldest son, was away at College, so he spent little time there. Steve, as well, was away, for the most part; first at Newberry College, and then West Virginia University, so the new household included my children Jacqui, Chris, and my wife, Karen. It was not a tranquil setting at the new house; the children and our housekeeper, resented my new wife's arrival and did their best to ignore her. Compounding, the family issues, my wife decided to fire our housekeeper, Delia, who had become part of my children's day to day life, setting the stage for an upcoming separation, and ultimately nasty divorce, in the early eighties. After my divorce from Karen, I continued living there for close to twenty-five years, until I sold the property in 2006 to John Marriott, the eldest son of Bill Marriott, who was at that time the heir apparent to his father, for assuming the Presidency of the Marriott Corporation. John immediately built a new home on the property, including building to house the some twenty antique and classic cars that his father had given him. I guess in looking back, that John qualified as one of the original "one per centers" in our population, before the "1%" divisor divided the country in 2011. Marriott was a great guy though, and I believe the other 99% would have liked him! My children also continued to live there, although shortly they would all be away at College, and one by one, graduate, and start their own marriages and families. My oldest son, Glenn, who graduated from Tulane University, in New Orleans, also continued there, finishing up with a law degree, and moving back to the Washington area, as an Assistant District Attorney in Maryland. Soon, after joining a law firm in Bethesda, Md., he married, briefly worked for me in the computer industry, and moved to Naples, Florida, with his wife and now, my two grandchildren, now working for his own law firm. Stephen, my second oldest, died in a motorcycle accident in May of 1980, just a little over a year after the death of his mom, Pat. A tragic loss and a loss that still today haunts our family, and emotionally, a hurt that never goes away. Chris, my youngest son, graduated from high school and attended the University of Vermont in Burlington, where he graduated and went to law school in Baltimore, graduating from the University of Baltimore Law School in the mid-eighties. He briefly worked in Maryland as an Assistant Public Defender, and from there joined me in my last Computer company venture, Computer Equity Corporation, where he ran a sizable portion of the operating company, until we sold it in 2001. Chris, happily married, still lives in Potomac, with his wife and two daughters, Kaitlin and Ashley. Jacqui, my youngest, moved on to school at Middlebury College, in Middlebury, Vermont, where she graduated with a dual degree in English and French, and followed with a graduate degree in English at the University of Toronto in

Canada. While she was attending school at Middlebury, I met Dr. Olin Robinson, who was then President of Middlebury, and we both became close friends, visiting and socializing together whenever we could. I would often criticize Olin for running a college, where kids would graduate from a highly ranked academic school, but have trouble finding employment after school. Who needs more English or French majors? One night, during dinner, when I was repeating this criticism, he looked up and said "it's ok to offer advice, Jack, but what can you do to help?" Stuck for an answer, by the time desert had been served, I had agreed to give the school $2 million, and we established the "Ballenger Computer Center", at Middlebury, in 1986. Over the years, the Center consisting of over 25 computer labs, offered a degree in computer sciences; and has graduated many a student, whom I'm sure found employment in the computer field, while at the same time being able to converse in French on their vacation to Paris! Jacqui, met a Canadian, while in school, a member of the Canadian Americas Cup sailing team; fell in love and married shortly after graduate school. They both moved to Middleburg, Va., adjacent to my farm, and raised two beautiful kids, Chloe and John, who are now attending High School in Washington, D.C. Sadly, my daughter died in 2009 at the young age of 47, becoming the third member of our immediate family to leave us. Jacqui suffered from multiple sclerosis for a number of years and finally succumbed to the disease, after battling it for a very long, painful period of time. There is no worst sorrow, than losing a child, but losing two, plus your wife has exacted its toll on me, as well as the remaining members of my family. I remember growing up, as a child, during the Second World War, when the newspapers were aglow, with articles about the Sullivan family that had sent five sons to war in the Philippines; and all five were killed in action, the entire family of children. How the parents of those kids must have felt!

With my children growing up and moving on to Colleges and marriages, my focus returned to C3 and guiding the Company's continued growth. At the end of our fiscal year in March 1982, the Company had posted revenues of $48.7 million and net profits of $8.4 million! Our stock price had moved to slightly over $50 per share and in June of that year, we split the stock on the basis of 2 for 1. The year 1982 also brought us a major contract win for initially $41 million that ultimately grew to over $75 million with the U.S.

Coast Guard installing workstations at Coast Guard Stations and Cutter ships on land and sea. We started a new commercial business called "Micro Products Company" which featured our own line of micro based workstations called the "Houdini" series used as a terminal device to mini-computers as well as an intelligent desktop computer that year; which we first demonstrated at the international computer conference held in Hannover Germany that summer. On a sad note, my good friend and fellow Director, Jack Veale died in June. Jack was a solid supporter of our Company, and his Company had helped me immensely in the first few weeks after I joined C3 in convincing the 3M company to continue buying our data entry systems. The other major news that year was that our Company was temporarily suspended from new contract work on Oct. 13, 1982, by the Defense Supply Agency with the Government claiming alleged false claims in billing procedures on our part. We quickly responded against what we were convinced, improper charges, and on October 27, 1982, the Government lifted the suspension, and our business continued as usual. However, this suspension placed an ugly cloud above the Company's head, and was a prelude to more trouble ahead.

Meanwhile, Vazzana was interested more and more in running day to day operations of the Company, so I began to search for a CFO to replace him, and found the ideal candidate in Frank Gaudette, an ex-Informatics employee that I had worked with at Informatics a few years back. John also knew Frank well and shortly he joined us in September 1982 as the new financial guy, eager to fill John's shoes! Frank, was originally from New Jersey; an ex-Army paratrooper, gruff outspoken personality, that everyone respected and warmed to. Frank stayed at C3 for close to 3 years before he was recruited by Bill Gates to become the first official CFO at a little firm named Microsoft, where he orchestrated the first public offering of Microsoft stock, and was one of the two or three key early players surrounding Gates, and was responsible for structuring a good bit of the Microsoft worldwide business plan growth. A charismatic guy; he married "Miss Minnie Mouse", who had performed that role for Disney, at Disneyland, Florida, and found time to be ranked in the top ten racket ball players, in his age group, on the East Coast. Frank was a guy you could never forget, once you had met him. He was a tremendous help to me in the

Figure 20: Listed on NY Stock Exchange From Left: David Schaumburg, Craig Sims, John Ballenger, Steve Hegeman, (NYSE Specialist), Marty Seldeen, Frank Gaudette

financial world, and I also taught him the "ways" of Wall Street, particularly involving him in the secondary C3 stock offering we did in 1983, raising another $16 million for the Company. That same year he assisted me in getting C3 listed on the New York Stock Exchange, a prestigious accomplishment for our firm. When Bill Gates went searching for a CFO, Gaudette was a perfect choice for Microsoft, who orchestrated their first IPO. When Frank later walked into my office to tell me that he was resigning, I remember the conversation like it was just yesterday. I said "Frank, you're an East Coast Guy, Why do you want to move West. Why? Your family is here; you've got a great job! Stay and I'll give you 6000 more shares of stock if you turn down the offer". The next day, he came in; told me he had thought about my offer, but was going to take a chance on the tiny company, Microsoft. He left in 1984, joined Microsoft, and five years later, his Microsoft holdings were valued at $5 billion! On a later visit to Microsoft, Frank introduced me to Bill Gates, and reminded me of the C3 counter stock offer, and after a few laughs, I left them both, thinking to myself, "I wish I had gone with him"!

I missed Frank when he left C3, because he really was a colorful character. At Microsoft, he continued his antics, as well as portraying his financial genius in a number of ways. When he took Microsoft public, he was heard to say, He was proud that he had negotiated an extra nickel out of the opening share price, when Microsoft did its IPO. At Company meetings I'm told, Frank always put on a show--being shot out of cannon, coming out as a boxer, etc. It was humorous as he was not a great public speaker, but when he spoke, one felt like they were listening to the Count from Sesame Street. He would just talk numbers "one billion, that's a big number,

I like that number". The classic Frank story that has been shared around the halls of Microsoft forever is that one day Steve Ballmer went in to talk to Frank, and the conversation got so heated that Steve punched a hole in Frank's wall. Frank followed Steve out into the hallway and said "you do that again and I will throw you out myself"! Frank's career at Microsoft ended just a few years after the Microsoft public offering. Unfortunately, he died of cancer at fifty-seven, at the pinnacle of his career.

By the end of 1982, Vazzana and I were also doing quite well with BALVA, the financial leasing Company we had started to provide lease financing for computer systems being rented by the Government. In the early 80s', the Government was more pro-active in leasing equipment rather than purchasing and since they normally would lease for a number of years, it became a lucrative business for BALVA. Additionally, it was advantageous for C3 to sell their computer leases for cash, thereby creating instantaneous bottom line profits for the Company; a plus for a public company interested in showing immediate profit growth. John and I had hired one of our earlier bankers, Joe Rockhill, convincing him to leave his bank and join us as the President of BALVA. A conservative banker, with outstanding connections in the banking community, he was not only able to manage the company, but with his banking relationships, found ways to borrow money to continue expanding our growing base of lease contracts.

It was the end of 1982, when I also decided to enter into the thoroughbred racing business with John, through Fletcher Jones at CSC. John had earlier began racing a sizable stable, with a friend, Vern Hosta, and I elected to follow suit, although on a more limited scale. By year's end we had purchased two horses, a Niginsky bred filly, we named DownStage for $325,000 and in partnership with a Kentuckian by the name of Price Headley, a mare, "Speedwriter", sired by Secretariat, for $300,000. Price was the son of Hal Price Headley, one of the 20th century's most successful thoroughbred horsemen, who owned Beaumont Farms in Lexington, Kentucky, a 2500 acre thoroughbred farm that had produced many winners over the years, including Menow, a horse than ran against Seabiscuit, several times, years back, and later the sire of Tom Fool and Capot, two Derby winners. DownStage turned out to be an instant winner and the two of us won purses worth over $500,000, before I sold out my half interest in Downstage for John's half interest in BALVA, worth around $500,000, in 1984. Shortly, after that sale, John was racing DownStage at Santa Anita track, when the owner of Seattle Slew made him an offer of $1 million; John sold, and Seattle Slew's owner retired Downstage, bred her to the "Slew" and the last we heard was selling Downstage foals for as much as a $1million a piece! Our other horse, in partnership with Headley, did not fare as well. Once purchased at the Keenland sales for $300,000, we immediately bred her to Riva Ridge, a famous Kentucky Derby winner for $25,000, and after delivering a beautiful colt months later, she refused to milk him and instead kicked the foal and killed him. Speedwriter did not last long in our partnership. Shortly after, we resold her in the Keenland, Kentucky Fall Sales. Through Price Headley, we were introduced to a number of famous horse buyers and trainers. On several visits to the Keenland Sales, I met and got to know Bunker Hunt, one of the wealthy and well known Hunts' from Texas oil money. Bunker was a true horseman; a gambler at heart with a string of famous thoroughbreds, and the individual who tried to "corner" the World's silver market in the 80s' only to go bankrupt several years later when that failed. A billionaire one day; bankrupt the next! Michael Sangster, another breeder, racer, and trainer who I also met at Keenland. Songster was a wealthy Irishman, who for years dominated the sales ring at Keenland, buying multi-million dollar race horses, until the Arabs, with their oil monies, replaced him in the 1990s' as the major purchaser of American thoroughbreds. My last horse purchased in the 1980s' was "Icebaby", a young two year old sired by the famous Kentucky Derby winner" Icecapade" for $75,000, whom I moved to Middleburg, Va., but never raced.

"Coming of Age"

When a young Company like C3 is coming of age, there are many exciting new horizons to set one's sights upon. It was a time of growth and change, as well as major and minor setbacks. Such was the case for C3. Revenue for the year ending March 1984 was $61 million and net profits of $4,300,000. The combination of a second Government suspension and investigation in January of 1984, over a contract dispute at the U.S. Army's White Sands Missile Range, caused a temporary stoppage in the award of new contracts to the Company, and also slowed delivery of systems under existing contracts. Thankfully, during the calendar year 1983, prior to the investigation notice, we had received a number of new contracts, increasing our backlog, and allowing us to continue shipments, while awaiting a decision on this second Army investigation in two years. Included in those 1982 contracts awarded were $5.4 million awards from the Dept. Of Labor and a new $19.4 million contract with the Navy over eight years for field maintenance services on existing C3 installations. The White Sands suspension also quickly was temporarily lifted, and shortly thereafter the Company received it's most significant contract award in March 1984, the last month of that fiscal year, a $70 million, ten year contract with the Navy for mini-computer integrated systems, that also turned out to be the largest single contract award in the Company's 16 year history. By year's end we had strengthened our branch operations to include over 65 domestic and overseas offices, including a new Southern Regional Office in Atlanta, and opened our first commercial International Headquarters located in Sunninghill, England. Our International network of Distributors of our commercial products stretched from the United Kingdom, Holland, Belgium all the way to Italy. Although elated by our continued contract award successes, the two Government suspensions and subsequent investigations were troubling to both the Company and to our public stockholders, and by year end, March, 1984, our stock price had plummeted to $6.75 a share.

There was never any doubt in my mind that these investigations had evolved as a result of the article Alan Abelson had written concerning our high profit margins, just prior to our going public. The investigations focused on profits and our purported over billing charges to the U.S. Army. All of our contract wins were fixed price contracts, competing against a multitude of other Companies, and since our bids were technically compliant and the lowest cost, we were able to increase our profits versus competition via of our integration methodology. In that era, there was a major Congressional effort underway to investigate and punish firms for fraud and waste, and the Government was looking for high level visible name recognition companies to attack. C3, although small, was a New York Stock Exchange Company, but lacked the financial size, strength, and political clout, of a company such as an IBM, to withstand the awesome resources of the Federal Government. So punishing us, a New York Stock Exchange Company with minimal financial capability to strike back, would be a publicity "feather in their cap" win for the Government investigators. Although the White Sands suspension had been lifted, based on information C3 had provided, the investigation continued and the Army refused to pay us for services and equipment that had been delivered. As a result we filed a lawsuit seeking back payments of close to $2 million owed us, followed by a countersuit from the Government seeking $9.4 million in false claims and penalties; and so the case continued into 1986 before finally being tried and settled in 1986. The stress and complications of our battles with the Government, spilled over into our personal lives, both at the Company and at home. Vazzana had separated from his wife Cheryl in 1983 and they divorced in 1984. I also had separated from my wife and our divorce proceedings were slowly working their way through the process with a final divorce decree occurring in 1986. Relationships between John and I also began to breakdown. The pressures and expenses of constant meetings with lawyers, and Government attorneys and procurement officials over the two investigations were heavy, and over time we found ourselves at odds over tactics and approaches to resolve the problems, and in late 1984, John resigned from C3, but elected to remain on the Board through 1986, and we decided to end our personal partnerships, specifically the thoroughbred horse and BALVA leasing businesses. The end of a 15 year management team that had become a "fairy tale" technology story, second to none! While all of this was underway, I was fortunate to be introduced on a blind date to a beautiful woman, Anne Brooks, who had recently moved east to the D.C. area from San Francisco. Anne was 36, recently divorced, and if you had met her then, you would have imagined you were in the presence of Natlie Wood, the movie star. She was a Natilie "look alike", and I was immediately attracted to her, but not her to me! After our first date, I asked her out again, only to be told, "sorry, but I don't think we have anything in common". Well, being a persuasive guy, we did go back out on that second date, and have now been together for some 28 years. John also remarried, a beautiful young gal, Tracey, in the mid-80s', who he was with until his untimely death in 2009. When we first met, Anne had taken a job in D.C. as the personal legal secretary and personal administrative assistant to Earl Silbert, a former prosecutor in the Nixon Watergate investigation, and lived in Georgetown until after my divorce in 1986. Anne was born in St Petersburg, Florida and lived there with her mother until her mother died when she was twelve years old. At that point in time Anne went to live with her Grandparents in Hendersonville, N. C. until her high school years, at which time she moved to Maryland with her father. After high school she moved to D.C. on her own; met a young man in the Army and married at 18. Her husband was then assigned to the Presidio Army Command in San Francisco, and shortly thereafter sent to Vietnam. While he was in Vietnam, she went to work as a secretary for several professional firms; and divorced her husband shortly after he returned from Vietnam. After ten years in California, working for several law firms as a legal secretary, Anne moved to Nashville, as a legal secretary for Harris Gilbert and then a politician, Jane Eskind, who ran against Howard Baker for the U.S. Senate. Anne then moved to D.C. where she met me. For me personally, my years with Anne have been the best since my life with Pat before her death in 1979. Many times, I have felt and sensed that Pat has orchestrated my 28 years of a beautiful relationship with Anne, culminating in our marriage in September 2011. I have been blessed and fortunate in finding Anne years back and equally fortunate in that my kids love her and are pleased that our relationship has been happy and compatible. This year my son Chris graduated from the University of Vermont and moved back to my home in Potomac. He had decided to follow a career in law, so the fall of 1984 found him entering the University of Baltimore Law School, commuting between Potomac and Baltimore. That summer after graduating from the University of Vermont, he worked for C3, as a summer intern, as did each of my children during their summer vacations. In previous years, both Glenn and Stephen worked overseas in our International Operations, reporting to Ed Spear, exposing them both to the workings of a business, coupled with the experience they gained in working in a foreign country. It was a wonderful feeling watching my children as they were growing up. I often thought of how my siblings had been separated, when my mother had died, and how great it was that I was able to keep our family together after Pat had died. We each had been through a mountain of hurt, but as the days and months had slipped by, we little by little were emerging from the years of chaos and sadness.

"1985-Renewed Growth"

Fiscal 1985 was a time of renewed growth and promise for C3. In July, 1984, the Company was awarded a huge contract by the General Services Administration for $73 million calling for the delivery of 4,299 office administration workstations over a period of eight years. Additionally, a contract with the Navy for $7 million was awarded providing up to 1500 Burroughs compatible terminals and printers. At the close of 1985, we had backlog contracts for over $209 million, at various government agencies, and commercial activities worldwide. C3 experienced a year of solid growth in FY 95. Total revenues for the year were $72.4 million, an increase of 18.6 % from revenues the previous year of $61.1 million, but profits fell dramatically to $6.8 million before unusual write-offs of $6 million. Our Tempest Division continued to grow, reporting revenues of $4.3 million and profits of $1.2 million. At the same time, our workforce had expanded to over 500 employees by year end. Over the last decade the Company had built a broad based field service organization which provided maintenance services for over 6,000 systems in 80 locations. The Company had now grown to the point where we occupied over 141,600 square feet of office and manufacturing space in Reston, and we were in the process of building a new 165,000 square feet corporate facility. Still lingering however, was the Army investigation relative to the Defense Supply Service contract. The hope was that this would be settled by the fall of 1985.

"Belt Tightening Times"

Our next year, fiscal 1986 was a "belt tightening" year for the Company. In addition to our continued problems with the Army and Justice Dept. Investigations, Congress had not yet approved a budget for the year, and Federal Agencies were forced to sharply curtail spending across the board. For C 3, this meant a slowdown in orders and shipments on existing contracts. Revenues dropped to $65.3 million for the year, and net profits to $1.5 million. The buildup of legal and lobbying expenses connected with the ongoing investigations was staggering, and before the issues with the Army were cleared in late 1986, we would incur over $5 million in legal costs defending the Company and Company officials from these charges. From the beginning, we had felt that the charges were unfounded, and fought settlement with the Government because we felt that we had been selectively picked by the Army and the Justice Departments to prosecute; to satisfy congressional committees intent on promoting their attack on fraud, waste, and abuse practices, thereby enhancing their image with the U.S. Public. In our case, what they were really attempting to do was to destroy a small company that truly was giving tremendous technology benefits to Government users for substantial cost savings, over larger companies charging more!

Despite lower than anticipated revenue for the year, we did accomplish a number of noteworthy accomplishments that year. Genicom, a printer manufacturer awarded us a commercial contract for $6 million to tempest proof their line of printer equipment and the Air Force Strategic Air Command awarded us a contract for approximately $20 million for tempest protected graphics terminal systems, and a follow on contract for $39.8 million of mini-computer systems for use in their AUTODIN classified network. This project was placed under the management of Ed Spear, who had transferred back from our International operations, and moved to Oklahoma City to manage the project. It was a huge contract that we knew would grow over time, and demanded a "top drawer" leader to supervise and control the program; so Ed was moved from our European Operations where he had done an outstanding job in organizing and building our overseas activities. Even with the cutbacks in Government spending, our backlog was continuing to grow from $209 million the previous year to a little over $218 million, as of March, 1986.

After more than five years, the Government's investigation of alleged false statements and false claims continued. On three separate occasions in Fy86 and FY 87, the Government filed for and was granted continuance running through September 1986. We would respond to charges, proving they were false, and the Government would side step the rebuttals, by requesting more time. It was obvious that their strategy was simply to wear us down; force us to incur additional legal expenses, and thwart our efforts to continue winning new contract business. The publicity being displayed in the press regarding this case was immense, and one can only guess that the Government had too much too lose without settling on their terms. This adverse publicity affected us dramatically. Frank Gaudette resigned to join Microsoft, replaced by Bill Gesell, who joined me from the General Electric Company, as Vice President of Finance and CFO. Several of our Directors left the Board, Allan Michaels of Convergent Technologies, retired Army Major General Jack Hancock, and Bill Thompson, formerly with Federal Data Corporation. In the meantime I had been able to recruit a seasoned competitor of C3, Richard Litzsinger to join our firm, as Executive Vice President, replacing John Vazzana. Dick was also elected to our Board, along with other replacements, Craig Sim, Senior Vice President of Donaldson, Lufkin and Jenerette, and John Woloszyn, a partner with the law firm of Frank Bernstein, Conway and Goldman.

As the year 1986 moved on and John and I continued to argue over the strategy we should pursue in this struggle with both the Army and now the Justice Department, it became quite apparent that the new Board had decided to “throw in the towel" and negotiate a settlement with the Government. So after several meetings between both, it was decided that the Company would agree to pay back all profits from the two contracts under investigation, along with penalties totaling $9 plus million, plus John's and my resignation from the Company for two years. If we had truly done something illegal, why would the Government agree to let us return after two years, or in the interim, even allow us to continue working in the Government contract arena period! Looking back on these times, it was the proper thing to do that is to settle even though we were found innocent. The past five years had been horrific for employees, stockholders, and management, specifically John and I. A lesson well learned by us was that there was really no way to have won this battle. The Government had invested to much in the way of time and publicity to admit they were wrong, whereas we had to cope with employees, careers and stockholder value. Litzsinger was promoted to President of C3 and John and I left the Company in the latter part of 1986.

"End of a Magnificent Dream"

C3 never regained its momentum, or significance in the integration marketplace after John and I left the Company. Litzsinger had been placed in my position to steady the Company and to seek a buyer, which he did. A little over a year later, the Company was sold in a leveraged buyout, and renamed "Telos" Corporation, which still exists today, located in Sterling, Virginia, run by the leveraged buyout group;, but sadly has had numerous troubles in paying off their massive LBO debt and finding ways to make a profit. When I often think back to comments made to me by others, that they liked the C3 story more than that of Apple, I dream of how it could have been; maybe not an Apple, who could match or top that story, but we had the team; the creative integration story, and the moxie to have become a multi- billion dollar leader in our industry. As it turned out however, all of our dreams were toppled by, in my opinion, a simple negative article written by Alan Abelson in Barron's newspaper!

Chapter 14 – Computer Equity Era

"The Come Back Kids"

Leaving C 3 after 17 years was indeed an emotional and devastating happening. To reach such lofty heights, only to see lifelong dreams and successes destroyed in a moment of time would normally be more than one could ever be expected to handle. But giving up was never written in the Ballenger or Vazzana Company handbook! John, now happily married to Tracey Servos, in 1987, purchased the Reston Country Club for $3.5 million embarking on a different kind of adventure, catering to the whims and egos of country club members, and running his stable of thoroughbreds around the country. Ed Spear resigned from C3 in 1986 and joined John in the Club adventure, running the day to day operations of the Country Club. After leaving C3, I decided to take off for a couple of months and spend time with Anne and my family, just relaxing. I had mentioned earlier, that when John and I decided to split up, I traded all my interests in the thoroughbreds we owned for his fifty percent ownership in BALVA, our leasing Company. BALVA, under the direction of Joe Rockhill, was extremely successful and profitable, leasing computers and peripheral equipment to the Government and Commercial customers, and as this business continued to grow, I decided after a few months to use some of BALVA profits to startup a new computer company.

At the time, BALVA was headquartered in a Virginia suburb called Chantilly, in a setting of small offices that Rockhill worked from; so using this office, I spent time over the next several months, laying out a new business plan for an area of the computer business just beginning to evolve. In the mid-80s' the computer industry had grown in an explosive way, and the emphasis was now shifting to the telecommunications arena. The need for moving large amount of data and voice precipitated the need for larger voice and data pipes and upgraded telecommunications lines. Also, old fashion telephony and PBX systems were being phased out, with new modern computer based telephone systems being the replacement choice. I realized that this market, although new, was similar to computer integration, in that we could integrate computerized telephone systems in almost the same way we built the integrated computer systems business over the last 18 years, and again, the Federal Government was an ideal market of choice. Overnight, in early 1987, I started the first of several new Companies that ultimately ended up under the umbrella of a holding company called Computer Equity Corporation (COMPEC). This first company I called Government Telecommunications Incorporated (GTI), and hired one of my original CDSI founder employees, Jim Collins, as President, using BALVA capital to fund it. As we did in the computer days, we began to meet and establish business relationships with vendors in the telephony world that manufactured computerized phone and PBX systems that could be integrated to meet customer requirements. Familiar names like Verizon, Sprint, Fujitsu, GTE, as well as many new companies in this field became suppliers to us, and it wasn't long before we became a major player in this new industry, and remained a major competitor in this business, until 2000, when Computer Equity and its subsidiaries were sold to a Florida a publicly company, Applied Digital Research Corporation for $39 million.

Over the next decade, GTI grew rapidly, winning major multimillion dollar contracts with large Government Agencies such as the U.S. Post Office, and the Social Security Service, automating their entire telephone, PBX, and telecommunications systems. It seemed there was no agency in the Government that did not use our integrated equipment. Major contracts with others quickly followed; the General Services Administration, our old customers the U. S. Army, the Air Force, and the Navy, as well. As we grew contract wise, so did our staff of people, and although never growing to the size of C3, when the Company was sold, we employed around 300 people nationwide, following the same C3 philosophy of maintaining a national maintenance staff to service what we sold. Further, the Government began to use our Company to upgrade their telecommunication cable and wiring, much of which was probably installed in the glory days of Alexander Graham Bell. Over the years, GTI, under the leadership of Jim Collins, and my youngest son, Christopher, who had decided to forgo a career in law for that of the telecommunications industry, installed several thousand telephony systems and several million miles of wire and cabling in various Government buildings country wide. Many of our key employees were people that had earlier worked for me at Computer Data Systems (CDSI) and C3; Kathleen McWilliams, my favorite classical pianist, Al Iwerson, who left C3 as Vice President of Operations, David Schaumberg, who had worked under Frank Gaudette at C3, joined as our first Vice President of Finance. Others like Tom Cuneo, one of the brightest computer technical marketing people in that industry, who had previously worked for me for over twenty years signed on, shortly followed by my oldest son, Glenn, who oversaw all of the legal and contractual aspects of our Companies. I mention "Companies" because as GTI grew under our umbrella Company, Computer Equity Corporation (COMPEC), we, one by one, acquired or started additional Companies, including a Dot-Com data mining Company we named E-Com Highways Inc., which I shall briefly talk about shortly. Crossing over from the world of computer integration to that of telecom integration was a smooth transition, since all of the new devices and systems were micro-computer built and software designed. It was a natural movement of our talents from one to the other, where we found ourselves competing against old line telephone Companies lacking much of the computer expertise that we enjoyed. Instead of competing against computer companies, we now were in a different world, up against new names; such as the Bell operating Companies that had been part of the old AT&T conglomerate, and had been broken up by the Government in the early 80s'. Entities like Bell South, Bell Atlantic, Southwestern Bell, Pacific Bell, GTE, Fujitsu; companies that had long legacies in the telecommunication industry, and like IBM in computers, enjoyed a firm "competitive lock" with their customer bases, until we came along with a new marketing and technical strategy, "Telecom Integration"!

Shortly after starting up Computer Equity in 1987, I happened to have a chance meeting with an old associate Fred Henschel, from my days at Computer Sciences, who was running a small trade show Company, Federal Contractors Corporation (FCC), owned by a group of investors in Philadelphia. FCC was a commercial company providing marketing support and equipment for a variety of trade shows they managed and setup for customers including computer and software companies. One of their largest accounts was with

American Online (AOL), at the time a small internet computer company started up and run by Steve Case. FCC had obtained the account with Case, and was responsible in this early days of AOL in setting up the AOL trade center booths at various computer conventions throughout the Country, including AOL computers, communications linkages, draperies, tables, phones, etc., prior to the AOL Staff arriving to man the show booth. Once a show was completed, they would dismantle the equipment and return it to their warehouse facilities in Virginia, awaiting the next show. COMPEC went on to acquire this Company for two reasons. One it gave our startup company some needed revenue and profitability, and two, more importantly it led us through AOL to additional telecommunication vendors and customers at trade shows that aided us in our main telecommunications business. It was interesting over the next few years to get to know a number of the early AOL employees from clerks to managers that became instant millionaires from their stock options held. One in particular I remember was "Willie" the number 3 employee of AOL, who was responsible for chauffeuring Case from place to place, and making sure his luggage arrived intact at his hotel. Willie's stock options turned out to be worth over $25 million, and Willie, a poor immigrant from the Philippines, retired from AOL and bought an island off the Coast of South America, where he resides today. Fred Henschel continued to run FCC, became a Vice President and Board Member of Computer Equity, and a valued business associate and personal friend of mine to this day. In 1998, we made our last acquisition. Henschel was close to a medium size trade show Company, "Expo Inc." headquartered in Kansas City, Kansas run by a talented woman, Carolyn Wiles, who had an excellent reputation in the trade show industry. We acquired this company in 1994, our second, with the idea of growing a chain of regional trade show companies. The trade show industry was, and still is today, a fragmented business made up of hundreds of "mom and pop" smaller companies around the country, and at that time, we were planning to build a chain of these smaller businesses by acquisition. Unfortunately, the idea died when we ourselves were acquired in 2000.

In parallel with my starting up and growing Computer Equity Corporation, John and Ed had developed and expanded the Reston Country Club in Virginia and decided to sell it to a Japanese investor in 1986 For $7.5 million. Always searching for a new venture, they shortly decided to buy a company, "Steelastic", in Akron, Ohio that made machinery for enhancing the manufacturing of steel belted radial tires. In 1990 after buying Steelastic, they acquired another Company NRM that manufactured equipment which made radial tires and merged the two together, now called NRM/Steelastic. With the merger of the two Ed moved to Ohio to manage the Company, which initially grew quickly with the tire manufacturing businesses expanding globally in the early 90s'. They were successful in selling equipment to a number of new startups in the tire industry including KUHMO, today one of the largest tire manufacturers in the world, located in South Korea. As part of this global explosion of tire Companies, NRM received a multi-million dollar contract from Iraq for equipment to startup Iraq's first tire manufacturing operation, just when Iraq decided to invade Kuwait. When that happened, President Bush froze all assets belonging to both Countries and unfortunately NRM-Steelastic was holding a multi-million dollar letter of credit drawn on the Bank of Kuwait that prevented NRM from collecting on the shipment. With the war underway, the equipment was lost in the "shuffle" and to this day, no one knows what happened to it. Confronted with this loss, the Company suffered an immediate huge cash drain, which John tried to plug by pledging all his assets, mainly his C3 wealth to save their Company, but unfortunately, it was not enough and the Company declared bankruptcy in 1992. John and Tracey had moved to Florida shortly after John had left C3, with John commuting back and forth to Ohio overseeing the tire company, with the demise of NRM, both Vazzana and Spear moved back to the Eastern Shore of Maryland. Disheartened by the loss of the tire business, it didn't stop them from buying Hitchler Industries, a company that manufactured plastic wood products out of recycled plastic bottles and cartons. They owned and ran this company for a few years, but back in the early 90s' plastic wood was still a few years away from being accepted by the public, so they made the decision to sell the Company to a new group of investors and both families, now a little larger with Gabriella, John and Tracey's daughter and Ed and Kim with their two, Ben and LeAnn, moved back to Northern Virginia. John shortly after joined up with a Company called Dunn Associates, as a Vice President; quickly became President of the Company, which he then took public and renamed "Steel Cloud", returning to our old line of business at C3 selling integrated computer systems to the U. S. Government. A year later, Ed Spear rejoined Vazzana, and when Vazzana left the Company, a while later, he became the next President of Steel Cloud. Ed stayed with Steel Cloud for a number of years, but ultimately retired, with Kim continuing to run her successful real estate Company in Northern Virginia. Both have built a beautiful home in Costa Rica and are planning a permanent move there in 2013. John moved on to becoming the President of a small biotech firm, Intralytics, headquartered in Baltimore, researching and developing treatment processes for a number of diseases, such as diabetes, and the elimination of contamination diseases such as salmonella, which he ran until his untimely death in 2009 of cancer. Tracey, John's wife, remained in Virginia, where she recently, graduated from Nursing School with high honors and is now working in Virginia for the Prince William Hospital System.

While GTI our telecom subsidiary was becoming a predominant integrator in the telecommunications market place in the late 1980s', the computer industry had changed dramatically. From mainframe to mini-computers to micro-processors, it had now become a low cost commodity PC based business, leaving little room for integration strategies to work. In essence profit margins had eroded, making it difficult for companies to grow, or better yet, survive. Further because the Government began to lease, rather than purchase PCs', the business became more risky, and lease contracts subject to cancellation, because of newer, faster and less costly systems constantly being announced by the manufacturers. However, knowing all of this, we decided to start a new subsidiary, Federal Systems Inc. (FSI) In 1991 to market "turnkey" systems, emphasizing added software, and or consulting value, in an attempt to set ourselves apart from competitors. I placed Kathleen McWilliams in charge of this Company as the President and for several years we struggled to compete, even won several multi-million dollar contracts, but after three years decided that it was just not a profitable business, and turned our new efforts toward the dot-com craze that took over the technology world in the mid-90s'. Kathleen was reassigned to our new dot-com startup, E-COM Highways, a data mining computer products entity, designed to be used by large computer commodity companies, providing instantaneous data online to their salesmen and tech support personnel from product description date, up to the moment pricing and availability data on their PC equipment, as they sat in a prospective buyer's office. This Company was also closed down shortly before we sold our parent company in 2001. We just couldn't sell the concept to the venture capital crowd! The concept was just not as exciting as many of the other dot-coms' that sprang up in the 90s', although a good percentage of them did not make it as well. I remember vividly one VC group we presented our business plan too at the end, criticized it as being too conservative! We were projecting profits at the end of two years in a slower, carefully executed way, but they were not interested in profits - "blowout" revenues to capture market share was the order of the day! Needless to say, tons of dot-Coms, with that mindset, disappeared in the ashes of the dot com bubble in the early 90s'.

We continued to spread our business activities across the country, and by 1995-96, COMPEC had become a household name in the hallways of Government Agencies and with vendors looking to market their products through our growing list of contracts. In 1993, we were selected by INC magazine as number 256 for their annual INC 500 fastest growing Companies in America. This would make the third time, two at C3 and once at COMPEC, that we would receive that honor. Local newspapers and journals, such as the Washington Post, the Washington Business Journal and the Washington Technology magazine, over the years, also selected us on a number of different occasions, as one of the Regions fastest growing Companies. In the late 1990s', I again began to talk with local investment houses about the possibility of taking COMPEC public. In my mid-sixties, by this time, I felt that a smaller regional IPO, or perhaps being acquired by a larger public Company, would fit within the timeframe of the number of years I wanted to continue working. My son, Glenn, had resigned, in early 1992, moving to Naples, Florida, where he reentered the legal profession, and my youngest son, Chris, along with other officers and Board Members, including Fred Henschel were not really interested in a long term commitment at the Company. The time was fast approaching where I felt it was in the interests of all COMPEC stockholders to look for a potential buyer, so after a few meetings and presentations to Ferris, Baker, Watts, and Legg Mason, the two largest regional investment

banking houses, in the Baltimore/Washington area; both firms began our search for an acquisition partner in the late 90s'.
By the end of 1988, after meeting with several companies interested in acquiring our firm, we selected a Florida Company, Applied Digital Systems, Inc (ADSX), a conglomerate of sorts, with varied businesses in the technology world. One of their subsidiary companies was "Verichip" a company that had created a computer system implanting chips in animals allowing vital storage information, ie identity data and medical information useful to owners, veterinarians ,etc. Verichip had also applied for approval from the U.S. government to implant these chips into human beings, thereby storing useful personal data for a variety of needs or uses.

The $39 million offer from Applied Digital was the highest received from potential birders with $10 million received at the time of the sale, with the balance to be paid based on a two year earn out period. As it turned out, the marriage of the two companies was not a happy one and after the culmination of two years of law suits between the two, we finally settled the case for approximately half of what we had originally sold the company for. It was a huge disappointment for the management and shareholders of COMPEC, and by the time the suit had settled, many of the key management and employees of the company had resigned.

As the Computer Equity story ended in the year 2001, so did my long career in the computer industry. After 46 years of working for, and starting up various companies, I decided it was time to leave that world to others and focus on the next chapter of my life.

Chapter 15 – Closures

It has all gone so fast! It seems like a brief flicker of time since I joined other earthlings in 1931, but truthfully all our lives are "brief flickers" and we have just a short span of time to make them meaningful and productive. Looking back, I think I have done that. Over the last 55 years, I have created personally, or with others, at least a dozen companies, of which 3 became public companies, employing ultimately some 5,000 employees. (See Below).

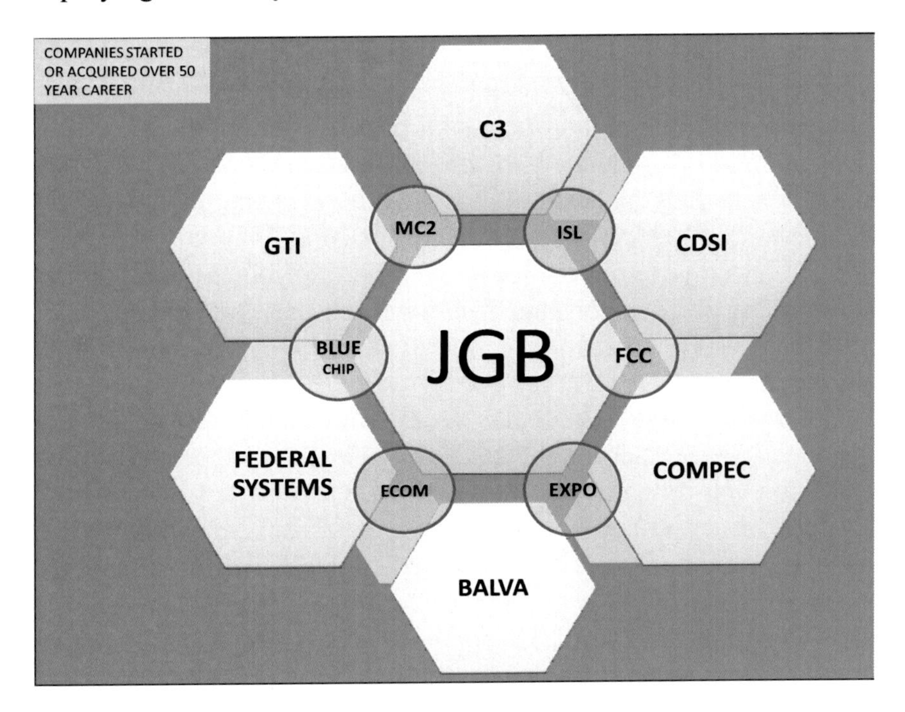

In the process of starting up and growing these organizations I have also created over 30 millionaires and multi-millionaires via stock and stock option ownership, over those years. Equally as important, I feel, is that over the years these companies have provided much needed quality state of the art services to both the private and Government industry marketplaces at completive prices, creating value and technical expertise thereby increasing the level of job opportunities as well as expanding the knowledge and experience of our customers and employees worldwide. My creating the systems integration industry ultimately spawned the evolution of over 200 companies in this business, providing new careers for tens of thousands of employees. There is no better feeling, than starting up a new venture, and watching it grow successfully. It is a certain "high" that only one that has done it himself, or herself, can truly understand and describe, but even an occasional failure, or setback is rewarding. Setbacks are a part of life and enrich our lives with additional knowledge and understanding that proves life offers "no free lunch", but new lessons to learn from disappointments and mistakes, As I have described in earlier chapters, there were certainly a number of misfortunes that sapped my spirit from time to time, but I was fortunate in using each to overcome adversity and continue on. It's what life expects from each of us! I have been truly fortunate in having both a successful business and family life. Of the two my family completely overshadows the former. I have been blessed with four beautiful children, and two beautiful wives, Pat, the mother of my children, with me for over 20 years, and Anne, who has been with me for some 28 years. Losing two children and their mother were major tragedies in my lifetime, but God has a way of Insuring that each of us is only given what we can handle, and I am thankful that He gave me Pat for twenty years and Stephen. And Jacqui twenty and forty-seven years respectively, before he took them. Watching each of the four grow into adulthood, has been a joy, and even though I lost Steve at 20, the years he spent with us was precious, and I shall cherish those years for the rest of my life. In an era where raising children is challenging and difficult, we were blessed in growing four "straight arrow" kids that any family would love to call their own. The same can be said for each of my six grandchildren, who are following in their parents footsteps. Jacqui's children, Chloe, in her senior year of high school at Madeira and John in his freshman year at Edmund Burke High in Washington, D.C., are both excelling in individual ways. Alexa and Glenn Jr. ("GJ"), my oldest son's children living in Florida, with Alexa are now in her first year of College, and "GJ", a junior in High School. The remaining two grandchildren, Kaitlin (14) and Ashley (8) daughters of my youngest son Chris are attending school in Maryland at Stoneridge and Holton Arms respectively. All six being the most attractive and intelligent human beings in the Universe! So says their Grandfather with righteous authority!

Perhaps the happiest part of my life over the past three decades was meeting Anne a long time ago. We have been together for more than 28 years; finally getting married this past September of 2011. I remember vividly calling Anne the morning after our first blind date, years ago, asking her out again, only to be told "Jack, I don't think so, we really don't have anything in common, so it's best you move on"! So, I guess it only took 28 years of "moving along" to find common ground before we got married, a long time just to make sure it was the right decision. Anne has been the anchor in our relationship over the years; a friend to my children; a beloved step mother to my grandchildren, and the centerpiece of our social life with our many friends and other family members. Maybe she was right 28 years ago about not having much in common with me, but the old adage "opposites are attracted to each other" has certainly been the case in our relationship.

Now in our retirement years, we spend our time traveling and visiting our many friends and family around the country, as well as entertaining them at our Farm in Middleburg, Virginia, and summer home in Rehoboth Beach, Delaware. Much of our time is spent managing and watching our thoroughbred racing operations up and down the East Coast. After selling Computer Equity Corporation in 2001, I decided to reenter the horse racing business, hiring Dale Capuano, as my trainer. Capuano, one of the top trainers along the East Coast, has served me well, and though I have not yet won any of the major stake races with him over the last ten years, we have been quite successful in winning over $600,000 in purses this past decade with more than 40 wins on various tracks from Florida to New York. In addition, Anne and I also both race together under the silks of B&B Racing Stable with her trainer, Susan Cooney. Susan is responsible for the training and racing of our horses in B&B, which we race primarily in Maryland, Virginia, West Virginia and Pennsylvania, We are both still looking for that "one in a million" to win the Kentucky Derby, but meanwhile we continue to have fun with our horse "babies", helping us both prove that we really do have a "little bit in common"!